# Intelligent Networks

# Intelligent Networks

Thomas Magedanz

and

Radu Popescu-Zeletin

INTERNATIONAL THOMSON COMPUTER PRESS

I(T)P An International Thomson Publishing Company

London • Bonn • Johannesburg • Madrid • Melbourne • Mexico City • New York • Paris
Singapore • Tokyo • Toronto • Albany, NY • Belmont, CA • Cincinnati, OH • Detroit, MI

Intelligent Networks

Copyright © International Thomson Computer Press

I(T)P  A division of International Thomson Publishing Inc.
The ITP logo is a trademark under licence.

For more information, contact:

International Thomson Computer Press
Berkshire House
168-173 High Holborn
London WC1V 7AA
UK

International Thomson Computer Press
20 Park Plaza
Suite 1001
Boston, MA 02116
USA

Imprints of International Thomson Publishing

International Thomson Publishing GmbH
Königswinterer Straße 418
53227 Bonn
Germany

International Thomson Publishing Asia
221 Henderson Road #05–10
Henderson Building
Singapore 0315

Thomas Nelson Australia
102 Dodds Street
South Melbourne, 3205
Victoria
Australia

International Thomson Publishing Japan
Hirakawacho Kyowa Building, 3F
2-2-1 Hirakawacho
Chiyoda-ku, 102 Tokyo
Japan

Nelson Canada
1120 Birchmount Road
Scarborough, Ontario
Canada M1K 5G4

International Thomson Editores
Campos Eliseos 385, Piso 7
Col. Polenco
11560 Mexico D. F. Mexico

International Thomson Publishing South Africa
PO Box 2459
Halfway House
1685 South Africa

International Thomson Publishing France
1, rue St. Georges
75 009 Paris
France

British Library Cataloguing-in-Publication Data
A catalogue record for this book is available from the British Library

Library of Congress Cataloging-in-Publication Data
A catalog record for this book is available from the Library of Congress

First Printed 1996

ISBN 1-85032-293-7

Cover Designed by Button Eventures
Printed in the UK by Clays Ltd, St Ives plc

# Contents

# Preface

Multimedia communications and global information connectivity are prerequisites for the society of the twenty-first century. Information and communication technology is a key factor in economic development, with computers and communication networks being the primary medium for all types of communication, including business, domestic, and entertainment. The reason for this is the convergence of the information technology, telecommunications, and entertainment industries; these were separate industries in the early 1990s, but the traditional boundaries are becoming blurred as digital technology becomes more predominant and demand for information services grows. This rapid increase in the demand for telecommunication services such as broadband, multimedia, and mobile and personal communication is a result of both technological innovation and other factors, such as political and economic changes.

The emergence of broadband network technologies over the past 10 years has provided the technological basis for the integration of different forms of communication. The trend toward integrated network technology and global network interconnectivity is the principal factor behind the emergence of a global network platform as a communication facility for all types of multimedia applications. The development of a computer-based information society in the foreseeable future requires appropriate communication services and information systems that are compatible with human communication processes. Time and distance are becoming less important in multimedia distributed applications as the network platform allows global multimedia interconnectivity. To achieve such a technical solution there are two main prerequisites:

1.  the development of end systems that are able to integrate data, audio, and video information in a way that matches human behavior and needs, the so-called 'mutimedia end systems';

2. architectures for multimedia distributed applications that model and support information flows in society in terms of a generic platform for open distributed environments.

These two areas of technical development are essential for the emergence of the so-called 'information society', 'information highways' or 'information market'; such terms attempt to capture the notions of globalization, autonomy, flexibility, scalability, heterogeneity, and mobility that are the characteristics of future distributed systems.

This book gives a snapshot of the technical development of architectures with adequate concepts for modeling autonomous, cooperative, and heterogeneous systems belonging to various operational, organizational, and technology domains.

The main topic considered in this book is the 'intelligent network (IN)' as one of the proposed and implemented solutions for such environments. We will provide an overview of the current state of the art in the area of INs, taking into account the historic evolution and future trends. In particular, we look at the motivation for INs, present the basic principles of the IN concept and provide a technical overview of what an IN is as it is widely understood today. This understanding is reflected in current international IN standards. We will therefore look in detail at the standards relating to aspects of IN service and architecture.

In addition, we consider the future of INs, taking into account the relationships between IN and other emerging concepts, such as management standards, broadband ISDN and mobile communications. Furthermore, we look beyond IN to the new information networking architectures, which are aimed at integration of IN, telecommunications network management (TMN), and open distributed processing (ODP) concepts and which are rapidly gaining momentum.

# About the authors

**Dr Thomas Magedanz** is Assistant Professor of the Department for Open Communication Systems of the Technical University of Berlin and his work focuses on distributed computing and telecommunications.

Dr Magedanz was born in December 1962. He received his masters and PhD in computer sciences from the Technical University of Berlin, Germany, in 1988 and 1993 respectively. Since 1989 he has been involved in several international research studies and projects related to intelligent networks, TMN, and TINA within EURESCOM and RACE as representative of Deutsche Telekom Berkom.

Since 1993 he has also been active in the fields of mobile computing and personal communications, leading a TINA-C auxiliary project studying personal communication support. Since 1995 he has been head of the 'Intelligent Communications Environments' Department of the GMD Research Center for Open Communication Systems (FOKUS) in Berlin.

Dr Magedanz is member of the ACM, IEEE and the GI. He is the author of more than 30 papers on IN/TMN integration and IN evolution.

**Prof. Dr Popescu-Zeletin** is Professor of Open Communication Systems at the Technical University of Berlin and Director of the GMD Research Center for Open Communication Systems (FOKUS) in Berlin.

Prof. Dr Popescu-Zeletin was born in February 1947. He graduated from the Polytechnic Institute of Bucharest, Romania, and obtained his PhD from the University of Bremen, Germany.

His professional activities have focused on computer networks, protocol development, distributed systems, and network technologies. Since 1986 his main research and development activities have concentrated on data communications in broadband ISDN, multimedia systems and platforms for distributed applications.

Prof. Dr Popescu-Zeletin is head of the department for scientific projects at Deutsche Telekom Berkom. Berkom aims to develop end systems, platforms, and integrated multimedia teleservices in a B-ISDN environment.

Prof. Popescu-Zeletin has served on various program committees of international congresses and conferences on broadband networking, management, multimedia systems, and distributed systems. He has published many papers on distributed computing systems, networks, and applications and has provided consultancy in data communications to national and international companies and organizations. He was active on standardization committees (DIN, ISO) and has contributed to the development of the ISO/OSI reference model and related standards. He is a member of the GI and the ACM, and a senior member of the IEEE and serves on the editorial boards of several professional journals. For his activities and results he has received the title of Doctor Honoris Causa from different technical universities.

# 1 Introduction

The *intelligent network* (IN) is currently the focus of worldwide attention. The IN is more than just a network architecture: it is a complete framework for the creation, provision and management of advanced communication services. The field of IN is becoming increasingly complex, but newcomers to the subject will find a gap in the literature. Most existing publications either tackle the topic from a high-level perspective or address specific aspects only. Furthermore, the standards are quite complex and thus difficult to read. Thus, it is often hard for novices in this field to obtain an overview of what the IN is and of related areas. We believe that there is a need for a tutorial on IN that provides both an easy introduction to IN basics as well as a comprehensive overview of the various concepts of the IN framework and its relationships.

This book is intended to constitute such a tutorial on intelligent networks. Its principal aims are to provide an overall understanding of the basic principles of IN and to present an overview of the current state of the art in the field of INs as it is widely understood today. This understanding is reflected in current international IN standards. IN standards are usually divided into the IN capability sets defined by the International Telecommunications Union (ITU) and the advanced intelligent network (AIN) releases defined by Bellcore. The ITU IN standards are more relevant for Europe, Asia, and Australia, whereas the AIN is more important in North America. In this book we concentrate on the ITU capability sets approach, as we believe that most of the ideas contained in this approach are also valid for AIN. This is particularly true when considering IN principles, the relationships between IN and other telecommunication systems, and the future evolution of IN.

This book is written for two kinds of readers who already have a basic understanding of the fields of distributed computing and telecommunications. The first category envisoned are those readers who require only a basic understanding of the IN and do not need to study the standards in detail. We suggest that readers falling into this category should read Chapter 2, which introduces the IN basics, and section 3.1.4, which describes some practical IN examples. Some parts of the other chapters may also be of interest, but these readers can consider Chapter 2 and section 3.1.4 as a stand-alone document containing all the information they need to know about IN. The second category of readers is represented by students and engineers who have to work with the IN and therefore need, in addition to a general overview, more detailed information in order to perform further studies. For these readers we have aimed to provide a detailed overview of the standards, with emphasis on an explanation of the interrelationships between the many concepts incorporated in the IN. Chapters 3 and 4 provide the level of detail required as a foundation for further studies. However, the information provided does not aim to replace the study of the IN standards. Rather, our tutorial should be regarded as a navigation guide through the available standards and additional publications. Thus, this book contains many references: we believe that it is worth consulting these references when working on specific aspects of INs.

It must be stressed that the IN as a framework for future telecommunications is a complex subject, as it incorporates many concepts; often the most difficult aspect for IN novices is the number of new terms with which they will be confronted. We have tried to keep this book as simple as possible, but we believe that is necessary to introduce and use the correct terminology in order to explain the concepts IN concepts and enable readers to understand and make use of the secondary literature. However, as aids, we have provided an index and glossary, which should support the easy retrieval of keywords and an explanation of their meaning respectively.

The book is organized as follows:

- Chapter 2 introduces the basic principles of the concept of the IN. We look at the business drivers for the IN and present the general IN objectives in section 2.1. Section 2.2 introduces the stakeholders that are present in an IN-structured network environment. In section 2.3 we look at the notion of services in the context of an IN, whereas section 2.4, which is the main part of this chapter, is devoted to the IN architecture. Here we introduce the IN architectural principles by illustrating the historic evolution of the telecommunications environment from the 'plain old telephone services' environment toward today's IN-structured environment. In this context we look also at the evolution of the main IN concepts and standards on a global scale. Finally, section 2.5 introduces the IN conceptual model, which is the current framework for the development of IN architectures.

- Chapter 3 constitutes the major part of this book and is devoted to the international IN standards. The main focus is on the ITU/ European Telecommunications Standard Institution (ETSI) standards. Section 3.1 considers the first set of IN standards, known as IN capability set 1 (CS-1). Based on the IN conceptual model we analyze in more detail the envisioned IN services and their realization within a distributed IN architecture. In addition, we briefly address service creation in section 3.1.3, and in section 3.1.4 we present four IN examples to illustrate service implementation within an IN-structured network environment. Section 3.2 presents the main enhancements of the CS-1 standards, addressed by IN capability set 2 (CS-2) and section 3.3 provides a brief look at future standardization activities. To take account the North American view of intelligent networks, we present a short overview of Bellcore's advanced intelligent network (AIN) in section 3.4. Finally, section 3.5 provides some information related to IN deployment and products.

- Chapter 4 addresses the future of INs, taking into account the relationships between IN and other major telecommunication systems, such as telecommunications management network standards, broadband ISDN, and mobile communications.

- Chapter 5 looks beyond IN to new information networking architectures aimed at the integration of IN, telecommunications network management (TMN) and open distributed processing (ODP) concepts, which are rapidly gaining momentum. Within this chapter we focus in section 5.1 on Bellcore's information networking architecture (INA) and in section 5.2 on the telecommunications information networking architecture (TINA) developed by the international TINA consortium.

- Chapter 6 provides a brief summary.

- The remainder of the book comprises references, a list of abbreviations and a glossary.

- Appendix A describe CS-1 services and service features, and also contains a mapping of service features onto services as well as a mapping of service-independent building blocks onto service features.

- Appendix B describes the CS-2 service features as far as they are known at the time of writing.

# 2 Intelligent network basics

The *intelligent network* (IN) is changing the telecommunication industry worldwide. The IN is a telecommunications network service control architecture that is a generic platform for open, distributed, service-independent communication. The goal of this service control architecture is to provide an open platform supporting the uniform creation, introduction, control, and management of services beyond the basic telephone services in the telecommunication environment. This is particularly desirable in an emerging deregulated telecommunications environment in which liberalization, and hence competition, is increasing, in conjunction with increasing customer demands for advanced communication services. The platform should be able to serve any number of users and a multitude of services with different attributes offered at various locations and by different service providers.

The IN architecture allows a variety of different services to be provided to customers independent of the underlying network technologies. The IN architecture defines a service-oriented functional architecture that enables the provision of a set of generic service components. These service components can be combined to construct new telecommunication services. Emerging services enabled by an IN architecture include flexible routing, charging and user interaction capabilities, as well as enhanced user control capabilities. Although most of the IN services promoted today can be achieved equally well individually within existing network environments, the major advantage of implementing a service by means of an IN is the uniformity of service creation, provision, and management, which results in greater economies and a reduction in the 'time to market'.

However, it must be stressed that the IN is more than just a new network architecture introducing a service-oriented network view: it is a complete framework for the uniform creation and provision of telecommunication services that will change the face of telecommunications fundamentally. In particular, the IN concept can be regarded as the ultimate step toward an integration of different network technologies, in which the IN provides a generic *application programming interface* (API) for network-transparent service provision. This means that IN is turning the 'network' into a programmable entity and thus paving the way for an emerging open market of (information) services.

In this chapter we provide an introduction to the intelligent network. To this end, in the next section we address the main IN business drivers and the basic objectives of the IN. In section 2.2 an introduction of the main stakeholders present in an IN environment is presented. Section 2.3 addresses the principles of IN services, which can be regarded as the main drivers for the IN architecture. In section 2.4, representing the main part of this chapter, we introduce the principles of the IN architecture, looking at the evolution from the traditional telephony environment, known as the 'plain old telephone service (POTS)' toward an IN-structured network environment. During this IN architecture introduction we also briefly consider the effects of the IN on the network platform, such as the signaling network. In addition, we provide an overview of the evolution of IN concepts and standards on a global scale. Finally, we introduce the intelligent network conceptual model, which is the worldwide accepted framework for the development of IN standards.

## 2.1 IN business drivers and objectives

The traditional telecommunications environment and the way in which services are provided to the customers within this environment means that we are facing a world of telecommunications monopolies, in which public telecommunications organizations act as both network operators and service providers. Consideration of the telecommunication services offered by these public network operators reveals that these services are driven mostly by available technologies rather than by customer demand. The reason for this is that the existing telecommunications infrastructure is mostly based on vendor-specific network equipment with proprietary interfaces, which is usually limited in its capabilities. Each network platform provides a specific, functionally limited set of services. There is no common methodology for the introduction of new services across multiple network platforms. Consequently, it takes usually a long time and requires huge investments to develop a new telecommunications service, as one network usually comprises equipment from several vendors which has to be updated (see section 2.4 for more details of this subject).

However, the telecommunications environment is currently undergoing dramatic changes. Some of the driving forces can be identified as follows.

- *Technological progress.* Progress in computing and telecommunications technology is proceeding at an increasingly fast pace. In particular, software will play a major role in future telecommunications. Based on the principles of distributed computing systems, the telecommunications environment is expected to move toward a distributed processing system in which the telecommunication services can be regarded as distributed applications (Chabernaud & Vilain, 1990), i.e. the network will become a programmable entity.
- *Deregulation.* The progression of deregulation on a global scale, driven by the concept known as open network architecture in North America (Basu, 1990) and open network provision (ONP) in Europe (EC, 1990), is reshaping the whole telecommunications sector. The existing monopolies will disappear as new players enter the scene, pushing for an *open* market of services. This market is not limited to specific country borders, but can be seen as a global market of services solving the communication needs of a global information society.
- *Customer demands.* The customer is playing an ever more demanding role in telecommunications; customers are no longer driven by available service offers but are playing an active role, asking for new and sophisticated telecommunication services to solve their business needs. In particular, the personalization of services, aiming for the provision of tailor-made telecommunication capabilities to everyone, anywhere, at any time, is an important market demand.

The establishment of an open market of information and telecommunication services, based on an open telecommunications environment, has resulted in a dramatic increase in competition. Probably the most important business requirement in such a competitive environment is to shorten considerably the 'time to market' of new telecommunication services. The main target is to reduce the time required to implement and deploy a new service from two or more years, as is presently the case with current technology, to only a few weeks or month.

Taking these requirements together, it can be appreciated that the traditional method of telecommunications service development and provision cannot meet the increasing demands of a competitive, highly dynamic telecommunications services market. Such a market requires an increased 'intelligence' of telecommunication networks, provided by 'software' whose importance is constantly increasing. This requires the introduction of new telecommunication service architectures that allow a new way of structuring of network hardware and service software.

This is where the IN plays an important role, as it is considered to be a technical solution to the above market needs. The IN is a key concept for the introduction of services beyond the basic telephone services in the telecommunications environment with shorter implementation times and lower costs than in traditional network environments because of its aim of unifying service creation, deployment, provision and management based on a service-oriented network architecture. In fact, the IN was developed in the US in the early 1980s to enable the rapid introduction of the '800 Service', which can be considered the most prominent IN service (see section 2.4 for more details).

The IN architecture approach has three major goals.

1. The IN architecture should be *service independent* to enable the realization of an open set of telecommunication services based on a common architecture. The functionality of the services should not be limited to the functionality of today's telecommunications services, e.g. voice telephony, but should support future customer demands, such as mobility, broadband, multimedia. The creation of new services should be possible easily and rapidly, as opposed to a platform defined for a specific set of services. This target is addressed by the IN through the definition of generic service building blocks, representing modular and reusable functions, that could be used for the construction of many different services.

2. The IN architecture should be *network independent* to allow the uniform implementation of services on top of any bearer network. This should eliminate the dependence of a specific network on the provision of service provider-defined services. This target is achieved through definition of service-oriented functional network elements, in which service switching functions are separated from service control functions. These functions could be flexibly allocated among physical network entities.

3. The IN architecture should support *vendor independence* to insure the interoperability of IN equipment provided by multiple vendors. This allows IN operating companies to combine multivendor IN equipment, as opposed to today's reality. This target is achieved through the definition of unique interfaces and protocols between the defined IN network elements.

Another basic objective of the IN architecture is to support a multiplayer telecommunications environment, in which a variety of services will be provided by multiple service providers in a competitive way. Thus, the IN has to provide appropriate interfaces to the IN platform in order to provide these service providers with access to the platform (Magedanz & Popescu-Zeletin, 1992).

Taking all these objectives into account, it becomes clear that the IN aims to provide a general platform for the uniform provision of current and future telecommunication services on a global scale. This global (IN) infrastructure will sit on top of existing and future (bearer) network technologies, combining the different methods of service provision based on the introduction of service-oriented network elements, enabling common service control principles. These new network elements will allow for the programming of the network in terms of service control.

## 2.2 IN roles

Taking the above considerations into account, it becomes clear that an IN-structured network environment is different from the traditional telecommunications environment. It is therefore essential to look at the stakeholders, which can be identified in an IN environment. Four stakeholders can be identified (Figure 2.1):

● network operator
● service provider
● service subscriber, and
● service user.

The *network operator* is a public or private company/organization that provides the IN-structured network and its resources for the execution of IN services. Note that the IN infrastructure comprises numerous network resources that have to be implemented above a specific physical (bearer)

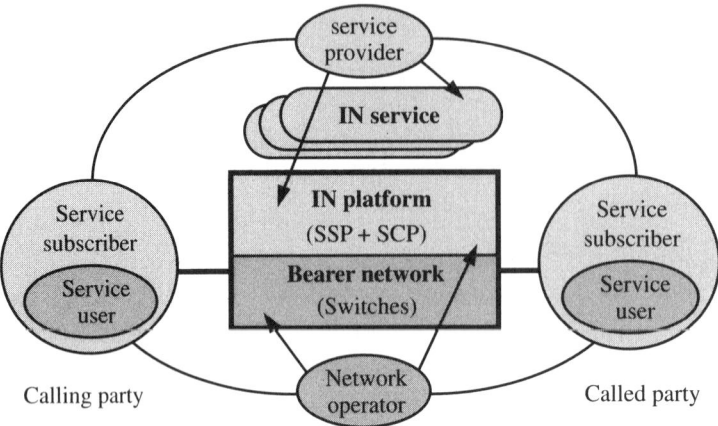

**Figure 2.1** Roles in the IN environment.

network, such as a the *public switched telephone network* (PSTN), *integrated services digital network* (ISDN) or even *broadband ISDN* (B-ISDN). The IN platform will allow service providers to supply IN services.

The *service provider* is a public or private company that develops and provides IN services commercially over the common IN-structured network and underlying basic (bearer) services. In specific cases the service provider may have its own IN network resources connected to the network operator's network infrastructure. The service provider is responsible for the provision and continuous availability of subscribed services. This means that its role is mainly in service management, comprising service creation and deployment, service administration (e.g. subscription), and service operation. Note that the service provider has to rely on the network operator for the execution of the services by means of the physical network and its resources.

The *service subscriber* is usually an organization that obtains an IN service from a service provider on a contractual basis and has to pay the charges to that service provider. However, residential users may also be service subscribers for some services, e.g. in the case of a call-forwarding service. Hence, the role of the service subscriber can be subdivided for some purposes into residential subscribers and business subscribers (e.g. a company), each having different service requirements. The service subscriber is responsible for operating its subscribed service, i.e. keeping it up to date for use according to needs. Thus, the service provider must provide administration functions to the service subscriber.

The *service user* is the person who has access to and makes use of a service, i.e. represents the called or calling party depending on the type of IN service, but will not necessarily be the service subscriber. For example, the service subscriber may be a company subscribing to a virtual private network (VPN) service and its employees may be the service users. A VPN is a service providing private network capabilities by using public network resources. The subscriber's lines, connected to different network switches, constitute a VPN, including capabilities such as a private numbering plan and call transfer. Other service users are the calling users of a Freephone service. Freephone, also known as 'Green Number' or 'Toll Free' service, is a service in which the called party, usually a large company, pays for the calls made to the Freephone number. However, service use is usually achieved by invoking a service (actually an instance of it) by calling a specific service or subscriber number.

It must be emphasized that the roles of the network operator and service provider may be adopted by the same organization, as is currently the case in most countries. However, continuing deregulation will result in future in the separation of these roles.

Note that the defined roles model two major aims in the definition and implementation of intelligent networks (Magedanz & Popescu-Zeletin, 1992):

1. a system allowing service providers to 'plug in' and offer new service easily;
2. a system that allows deregulation by providing platforms which offer functionalities at the same price for all in a competitive deregulated environment.

For completeness it must be stated that, in addition to these basic roles, other roles exist. These are, for example, external service designers, manufacturers, and vendors of telecommunications equipment. However, these are less important in the context of this book.

## 2.3 Services in the IN context

The term 'service' is quite confusing at present as it is defined and understood differently by the different providers and users at different levels in the IN architecture. In what follows we will attempt to classify and define the notion of service in the IN context. The notion of service was initially introduced by the ISO/OSI Reference Model as one of the two necessary concepts to insure layer independence: the *service* and the *protocol* concept. A service in this context is defined as the set of capabilities offered by a layer (platform) to the layer

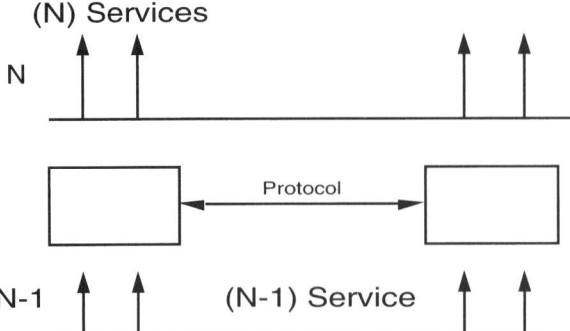

**Figure 2.2** Relationship between services and protocols within OSI.

above and is independent on any particular implementation. It expresses the external functional visibility of a layer (based on the functionality of the layer below; Figure 2.2.)

A protocol defines the rules of interaction between entities that are situated in the same layer (platform) but which pertain to different systems in order to provide a certain layer service. It is obvious that the definition of the service offered by a layer/platform and the definition of the service offered by a lower layer completely specifies the functionality of a layer (platform). Different protocols may carry out this functionality. From this point of view we may identify in the IN architecture:

- services offered by the underlying network platform, which are known as 'bearer services', such as audio, video, or data transmission and signaling services; and
- services offered by the IN platform to the end users, i.e. service subscribers, representing enhancements of these underlying services and sometimes referred to as 'supplementary services' or 'value-added services'.

In this book we want to concentrate on the second category. One major aim of the IN is to meet increasing customer demands for advanced telecommunication services. By looking at the basic attributes of emerging telecommunication services it is possible to identify service elements that are common to several services. An IN network provides the elements necessary to support these generic functional characteristics common to IN services, such as flexible routing, flexible charging, advanced user interaction, and enhanced customer control over specific service elements.

It should be pointed out that it is not within the scope of IN to define any specific IN services, as the IN aims to provide a service-independent network platform. In order to support the provision of an open set of services, the IN defines an extensible set of generic and reusable service components, known as 'service-independent building blocks (SIBs)'. These can be combined into higher level service elements, known as 'service features', to enable the rapid construction of new services within the scope of the service components' capabilities. This allows service providers to respond quickly to specific market demands. By evolutionary extension of the service components' functionalities (and corresponding enhancement of the IN architecture), more advanced telecommunication services can be achieved, while still supporting existing services. Although most currently promoted IN services are related to advanced telephony, i.e. voice applications, future IN services will also include mobility, broadband, and multimedia applications, i.e. the IN concept will provide the necessary openness to enable the realization of future services.

In fact, two main classes of IN service elements can be identified. (For completeness we should mention that recent IN standards also include the notion of a third category of service elements, referred to as 'service creation services', but this kind of services has not been studied in detail so far.)

- *call-related service elements*, such as call forwarding, call screening, and time-dependent routing (i.e. 'intelligent' number translation services in general); and
- *management service elements* related to service and customer management, such as billing, statistics, and service customization.

Examples of IN services defined and implemented by means of reusable service components are Freephone, televoting, premium rate, card calling, virtual private networks, and universal personal telecommunications. A list of envisioned IN services, along with their description and decomposition into service features and SIBs, is given in Appendix A. In addition, section 3.1.4 provides a more detailed illustration of the realization of the first four of these IN services within an IN architecture. Note that the users of an IN service do not know or care how the service is achieved in the platform, but only how it is provided to them. In other words, any of these envisioned IN services can also be implemented in a conventional network, but the IN-based service implementation is the most economic method.

## 2.4 IN architecture

In this section we look at how the traditional telephone network evolved toward an IN-structured network environment. In particular, we consider the traditional method of telephony service provision, explain the basic idea of the IN approach, look briefly at the technological changes that have enabled the introduction of INs, i.e. the introduction of outband signaling systems, provide an overview of the evolution of IN concepts, and finally present the IN conceptual model, which is the general framework for IN development today.

### 2.4.1 The traditional POTS environment

The first 90 years of telephony was devoted to the global provision of the *'plain old telephone service' (POTS)*. As a result of the introduction of computers in the telephone network for switching control, a dramatic change took place in 1965. The first 'stored program control' switching systems were integrated in the network, allowing for additional *switch-based* services, usually referred to as 'supplementary services'. Examples of such advanced services are 'call forwarding' and 'abbreviated dialing'.

Traditional public networks, such as the *public switched telephone network* (PSTN), are usually composed of heterogeneous switching systems from different vendors and with different technologies. The lifetime of these systems is long (e.g. more than 20 years) because they represent a substantial investment for the network operators. They are also quite complex to implement; usually the software size is of the order of more than one million high-level programming language lines. Consequently, the introduction of new telecommunication services is a complex process in such an environment (Figure 2.3).

In the traditional POTS environment the switching systems, typically referred to as 'switches', 'exchanges', or 'offices', perform the basic call processing, i.e. each switch contains a so-called 'basic call process', which supports the provision of a particular basic (telephony) service. Each supplementary service represents a monolithic, non-reusable software entity that modifies or supplements the basic call process. This means that the introduction of a new supplementary service requires the modification of the basic call process (in the switches). Furthermore, the switching systems have to provide, in addition to their (basic) switching capabilities, translation tables (i.e. the service data) and corresponding service programs in order to provide these supplementary services.

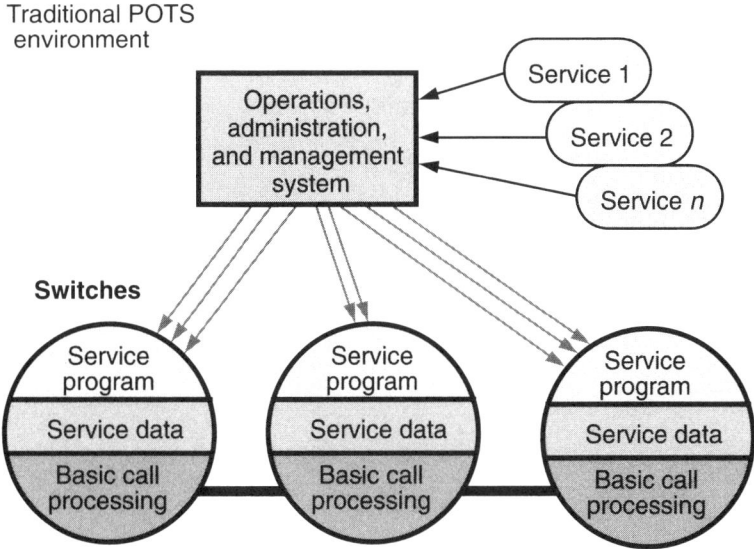

**Figure 2.3** Switched-based service provision in the traditional POTS environment.

A switching network usually consists of a hierarchy of switches, e.g. local exchange level, intermediate exchange level and transit exchange level, as indicated in Figure 2.4. Switch-based service systems are usually introduced at the transit level so that they are available to larger geographic areas. Although this approach works well for limited service use, it becomes inadequate for high service use because of the resources (number of switches and the related trunks) required in processing the switching-based service. Thus, service data and programs are migrating to lower levels of the switching hierarchy (e.g. intermediate or local exchange level), enabling better resource utilization and performance for high service use. In the most extreme case, each switch (at local exchange level) contains this data as depicted in Figure 2.3. This means that *every* service must be loaded into *every* switch's software before it can be used!

It must be stressed that with the increasing number of switches hosting a particular service the introduction and maintenance of services becomes increasingly complex. Hence, it is difficult to maintain consistently the service data, i.e. the corresponding translation tables, replicated in all network nodes, on a large geographical scale.

In addition, the introduction of new features and services in a multivendor environment is an expensive task. In particular, as the number of network

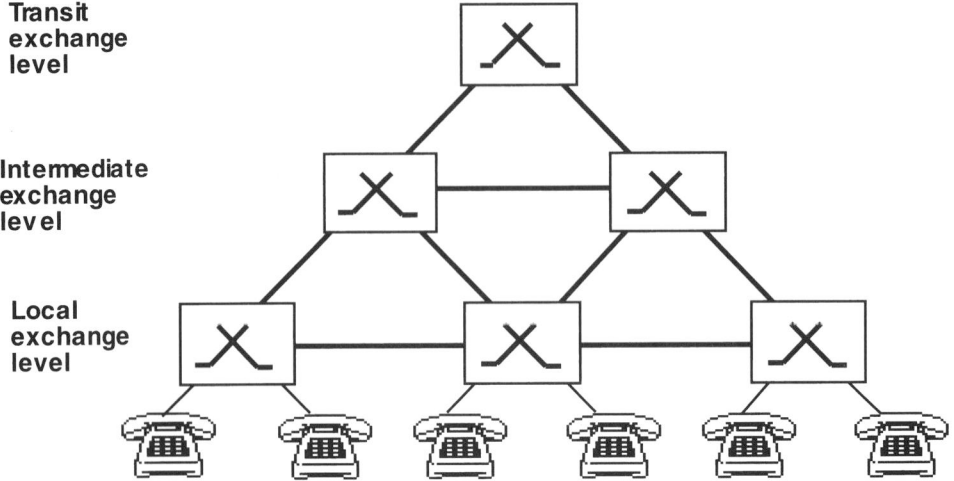

**Transit exchange level**

**Intermediate exchange level**

**Local exchange level**

**Figure 2.4** Levels of a switching hierarchy.

services increases, it becomes increasingly difficult to introduce new services in the next system software release (typically, there is a new release every 1 or 2 years). As a consequence, it usually takes several years (usually 3–4) to introduce a new service.

Figure 2.5 depicts an example of a non-IN service implementation of a Freephone service, known as 'Service 130', in Germany. A service user dials a (service) number beginning with the digits '130', identifying the Freephone service. The local exchange performs normal translation and recognizes that it is not able to route the call to the final destination, since it considers '130' to be another location area code. Thus, the local exchange routes the call by establishing a (bearer) connection to a switch higher in the network hierarchy. By this procedure the call will be routed to an appropriate switch, usually a transit exchange, containing the 'Service 130' database and service program.

The transit exchange, containing the service logic and data, recognizes that it is able to translate the logical '130' number into the appropriate physical destination number. After performing the number translation, it routes the call through the network to the subscriber being called, e.g. '25499-229'. In addition, the transit exchange establishes the specific charging procedures in order to charge the called party for that call.

As indicated in Figure 2.5 this method is costly because of the resources (i.e. lines and exchanges) involved in the call processing. Thus, in large networks several higher level 'Service 130' exchanges are required to support this service

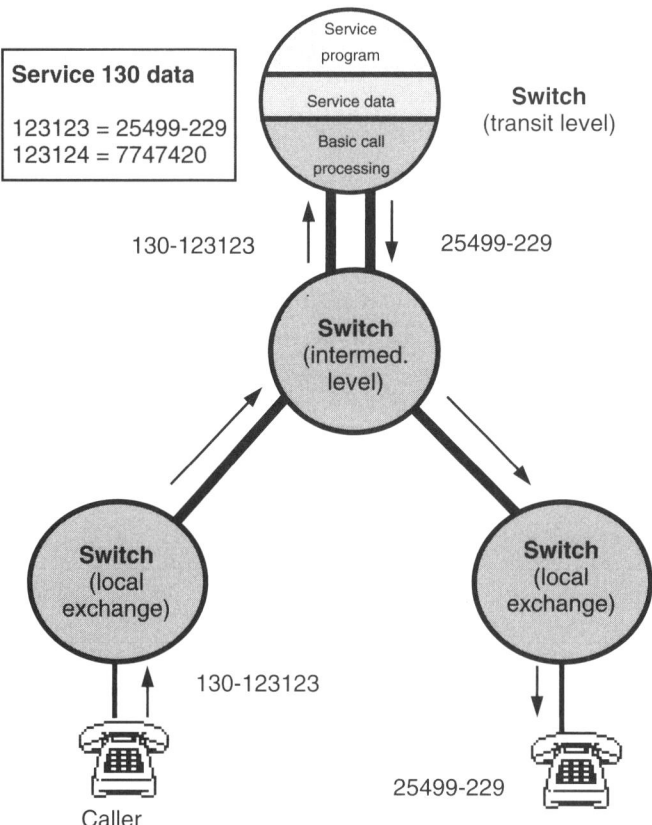

**Figure 2.5** Non-IN Freephone service implementation.

on a nationwide basis and improve the efficiency. However, this in turn makes updates and the administration of the service complex.

## 2.4.2 Toward the first intelligent network

In the beginning IN technology was motivated by the need to provide additional capabilities and control in the plain telephony network. The plain telephone network was a pure switching facility, and was not able to provide new routing, charging, and number translation capabilities. There was a need for greater 'intelligence' inside the network to support additional functionality.

The first step toward solving the above problems was to keep the service data necessary for a service outside the switching network in a centralized database accessible to the different switching nodes. All vendor switching systems were required to be able to access the database via a standardized

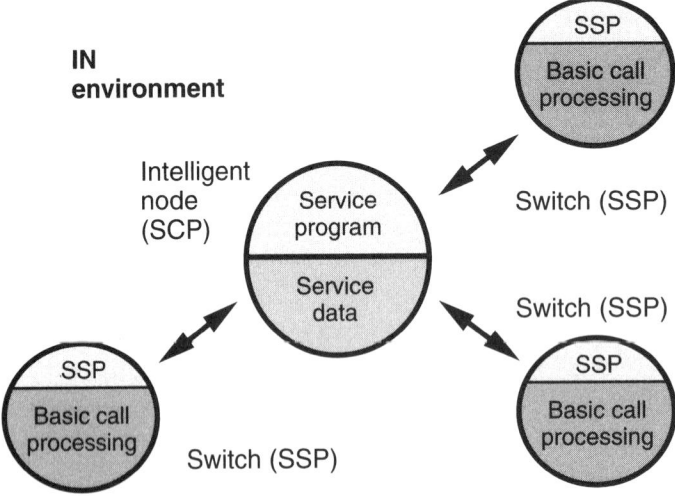

**Figure 2.6** IN-based approach.

protocol. Very quickly this approach was not seen to be sufficient with regard to the service programs (known as 'service logic' in IN terminology) for upcoming services.

The next step was to outsource also the service logic and to define a switching system-independent interface/protocol to 'intelligent nodes' containing both service logic and data (see Figure 2.6). The advantage of this approach is obvious: by putting a new service program and the related service data outside the switching network, it is possible to introduce a new service ubiquitously throughout a given network. A flexible scaling of the service at different hierarchy levels is another advantage of this approach. In addition, this new network architecture allows for more competition in the provision of telecommunication services than traditional network environments (Magedanz & Popescu-Zeletin, 1992).

However, the fundamental prerequisite for this approach is the 'real-time' connection between the switches, referred to as *service switching points* (SSPs), and the 'intelligent nodes', known as *service control points* (SCPs). This fast, reliable, and standardized interconnection of service switching points and service control points forms the basis of the IN architecture and became possible through the introduction of 'common channel signaling' systems based on the new concept of 'outband signaling' in the 1970s. The idea of *common channel signaling* is a simple one. Unlike the traditional method of 'inband signaling' in which both call/connection control information and user information are exchanged via the same data channel, inter-switch common channel signaling is based on digitally encoded messages carried in a *separate* signaling 'trunk', which allows for faster and more reliable call/connection control. ['Signaling' is commonly understood as the ability to set up, control, and release (bearer) connections.]

The first common channel signaling system introduced was AT&T's common channel interoffice signaling network, followed by ITU's signaling system no. 6. In the 1980s ITU defined the *common channel signaling system no. 7* (CCS7), which today forms the heart of modern telecommunication networks. CCS7 is a logical, separate 'out-of-band' signaling network used for the control of the bearer service network. CCS7 could be considered to be the central nervous system of telecommunication networks and is the foundation for a large number of applications in telecommunications networks, such as ISDN, mobile networks, and IN.

Relying on the common channel signaling network for the provision of IN services makes it possible to place the service logic and data into dedicated nodes in the network that remotely control the establishment of call connections at the request of the switches. Hence, services can be easily introduced or modified in an intelligent node. Note that within this approach service switching and service control have been separated, requiring the definition of the corresponding 'interfaces' between the switches and the intelligent nodes.

In this context two basic items have to be addressed. First, the basic call model in the switches has to be enhanced to provide for the flexible provision of basic and supplementary services. Thus, 'hooks' identifying the trigger events in the switch have to be added to the basic call process to start interactions with the external IN service logic. Secondly, a corresponding protocol has to be defined for the dialogue between the service switching point and service control point on top of the signaling network.

The early IN architectures were service specific, i.e. only designed to support, for example, the '800 Service' in the US (see also section 2.4.4). These architectures made use of specific call models and specific signaling network protocol capabilities, but the supported message sets were tightly coupled to the specific services.

In order to support multiple services by the IN architecture, both the call model and the protocol used between the switch and the central control node should be service independent. The basic point of this evolution step is the definition of a generic interface. Hence, IN architectures decouple IN capabilities from specific services by providing generic message sets and switch call models that are applicable to different IN services. These generic

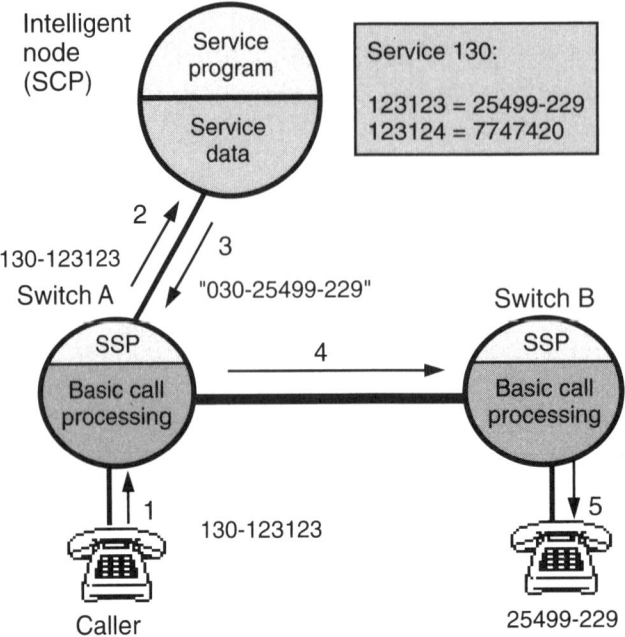

**Figure 2.7**   IN-based Freephone service implementation.

message sets and call models apply at both originating and terminating ends of the call.

Figure 2.7 depicts an example of the IN-based implementation of a Freephone service that allows users to call a specific number toll free, since the called party will pay for these calls (e.g. the '800 Service' in the US or 'Service 130' in Germany). In this service scenario a service user dials the Freephone service number, which starts with '130' as Freephone specific 'service access number'. (Each IN service number consists of a service access code/number and a service subscriber-specific number. In our scenario '130' is the service access number and '123123' is the Freephone subscriber number.) The local exchange, i.e. the service switching point, recognizes from this service access code that this call is an IN call and queries the service control point for call-handling support. The service control point checks its service database and translates the logical '130-123123' service number into the corresponding destination number, for example '25499-229'. This information is passed back to the service switching point in order to resume call handling. With this new information, the service switching point is able to set up the call, i.e. to establish the corresponding bearer connection to the appropriate final destination.

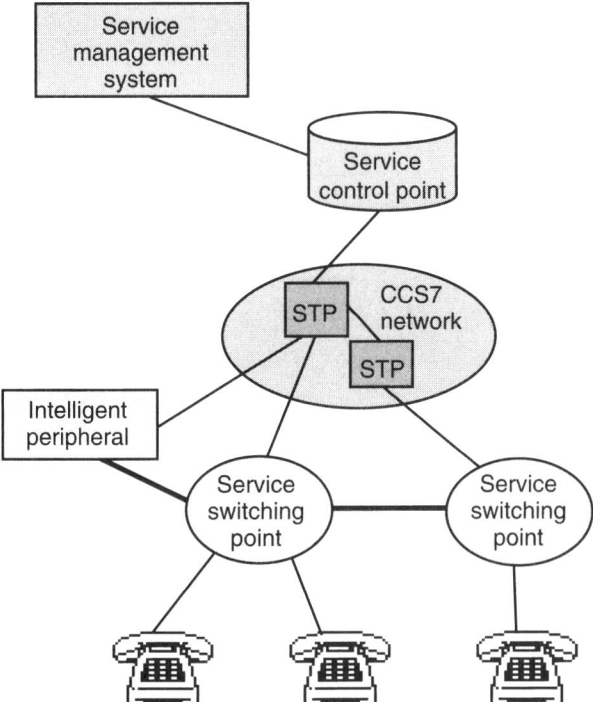

**Figure 2.8** Basic elements of an intelligent network.

Note that the SSP–SCP communication will be achieved via the CCS7 signaling network and not via the bearer network.

After these initial considerations it seems logical to look in more detail at the basic network elements of an IN architecture. The following network elements can be distinguished in every IN architecture (as depicted in Figure 2.8).

SSPs are stored-program control switches, either local exchanges or access tandem exchanges, that are able to interface with the CCS7 signaling network. These switches also contain a limited 'service (access) logic' required to suspend calls that require special handling. The difference between today's traditional exchanges and the SSP is that the reaction of the SSP to customers' requests for a particular service (this means that a user has dialed an IN service number) will be controlled by a regional SCP, rather than directly by the local switching exchange. The SSP recognizes IN service calls and routes the corresponding queries to the SCP via the CCS7 signaling network, which consists of *signaling transfer* points (STPs). SCP commands will be used by the SSP to process the call further. Note that a SSP can address several SCPs.

The SCP is typically an on-line, fault-tolerant, transaction-processing database that provides call-handling information in response to SSP queries. SCPs are high-capacity systems handling several hundred transactions per second or more than 100 000 calls in an hour. For a mated SCP pair the response time requirement is less than half a second and downtime should be less than 3 minutes per year. An SCP has several signaling network interfaces. The SCP is designed to accommodate growth, which means that processing power or memory can be added to an in-service SCP without interrupting service handling. SCPs are designed to support multiservice operation.

In order to manage IN platforms, a *service management system* (SMS) containing the reference service databases is required. Supervision, remote operations and maintenance of SCPs, and (coordinated) software downloading are some SMS features. Transactions from the SCP to the SMS (usually in 30-minute intervals) include performance measurements, traffic data, and billing data. Both network operators and customer terminals or computers can communicate with the SMS to retrieve service reports or update data. The SMS is integrated in an operations support system that supports network operation, administration, and maintenance functions, and normally resides in a commercial host computer.

An additional *intelligent peripheral* (IP) may be connected to an SSP, providing enhanced services/functions, such as announcements, speech synthesizing, and speech recognition, under the control of an SSP or SCP. The motivation for the introduction of this network element is financial, because it will be cheaper for several users to share an IP when the IP is too expensive to be implemented at all SSPs. IP functionality is usually required for interactive, e.g. voice-prompted, IN services.

Note that the introduction of an IN architecture in the public network environment is a great challenge owing to the flexibility provided for the introduction of new services and by creating the technological basis for competitive service providers. The major cost is incurred by SSP introduction, as this IN element has to be introduced widely and requires CCS7 availability for the switch. Typically, the introduction of SSPs starts at a high level in the switching hierarchy, e.g. at the transit exchange level, and migrates down to the local exchange level when IN service call volume increases.

## 2.4.3 IN impacts on the network platform – CCS7

As mentioned above, the introduction of 'outband signaling' based on ITU's *common channel signaling system no. 7* (CCS7) is the fundamental prerequisite

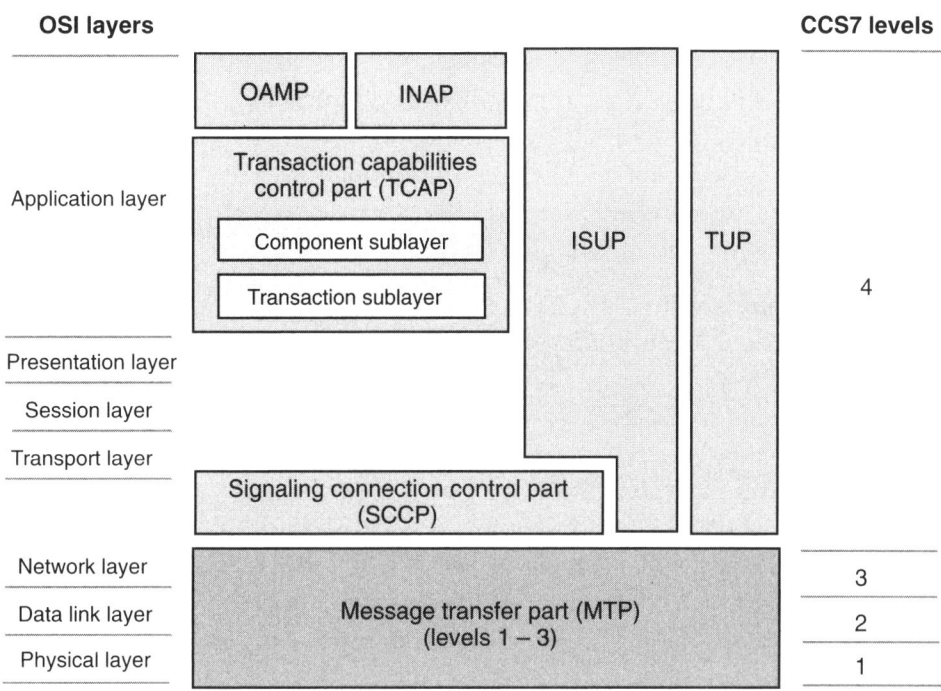

**Figure 2.9**  CCS7 protocol architecture.

for the IN architecture, connecting the geographically dispersed islands of intelligence. Besides SSPs and SCPs (acting as 'signaling end points') the CCS7 packet network encompasses *signaling transfer points* (STPs), as illustrated in Figure 2.8. STPs are very high-capacity, very reliable packet switches that transport signaling messages between the network nodes. To perform these functions, they need a large routing database containing translation data. An STP typically terminates 256–1024 signaling links, processes 1000–5000 signaling messages per second, and introduces packet delays of less than 100 ms. The allowed downtime from one signaling end point to another should be less than 10 minutes per year.

The CCS7 protocol architecture is adapted to the seven layers of the ISO/opens system interconnection (OSI) reference model (Mitra, 1991) (Figure 2.9). The functionality of CCS7 can be mapped to the OSI layers. A common signaling transport capability referred to as *message transfer part* (MTP) covers OSI layers 1–3. Several other applications, covering the upper OSI layers are referred to as 'user parts' or 'application parts'. The defined application parts are:

- telephone user part (TUP) for basic telephony;
- data user part (DUP) for circuit-switched data services;
- ISDN user part (ISUP) for combined data, voice, and video services;
- mobile application part (MAP) for mobile telephony, such as within the global system for mobile communications (GSM);
- operations, administrations, and maintenance part (OAMP); and
- IN application part (INAP) for IN services.

The MTP comprises three (CCS7) levels, which collectively provide a highly reliable and resilient connectionless message transport mechanism. Level 4 is partly covered by the *signaling connection control part* (SCCP), enhancing the limited capabilities of the MTP to provide a connectionless (i.e. datagram) as well as a connection-oriented message transport. In addition, the SCCP enables enhanced addressing capabilities for routing messages within the CCS7 network and further distribution within its network node.

The *transaction capabilities application part* (TCAP) provides application layer formatting and procedures for real-time intensive query/response-type applications. A typical example of a real-time application using TCAP is a number translation function, in which an exchange queries a remote database to perform the remote operation of converting a specific service number (e.g. '800-xxxxxx') into a network-routable number.

TCAP is composed of two sublayers, the *component sublayer* and the *transaction sublayer*. Whereas the component sublayer is modeled after the OSI *remote operations service element (ROSE)* protocol, the transaction sublayer has no equivalent standard in OSI. It is designed to set up and terminate an end-to-end 'association' or 'connection' for application protocol exchanges over the connectionless SCCP service. The transaction sublayer is used to group components (i.e. remote operation invocations and their associated responses) within the context of a 'dialogue', also known as 'TCAP transaction', between two application entities.

It is this functionality of TCAP that provides the basis for many application parts, and in particular the INAP ((ITU Recommendation Q.1218; ETSI ETS 300374-1). INAP defines a set of operations and the corresponding protocol required for the dialogue between the IN elements, such as between SSPs and the SCP.

More information on CCS7 can be found in Modaressi & Skoog (1990). More information on INAP will be given in Chapter 3 (pp. 71–4).

## 2.4.4 From the first IN toward today's IN

Several IN architectures that have concepts in common but differ in their scope, terminology and their means of description have been defined or are under development, as illustrated in Figure 2.10.

The development of INs started 30 years ago with the introduction of outband signaling systems. The first IN-like architecture was proposed by AT&T in the US in 1976 based on *'stored-program-controlled'* switching systems. This proprietary architecture was based on the definition of a so-called 'action control point' and a 'network control point', allowing remote access to centralized services.

However, the pioneering work on INs has mostly been done by Bell Communications Research (Bellcore) in the US with the proposition of the *intelligent network 1 (IN-1)* (Ambrosch *et al.*, 1989) and *intelligent network 2 (IN-2)* (Bellcore, 1986). IN-1 was originally designed for the provision of the '800 Service' (known internationally as the 'Freephone' or 'Green Number' service) and introduced first by Bell South in 1987. The major difference between IN-1 and IN-2 was that the latter was intended to be service independent by the

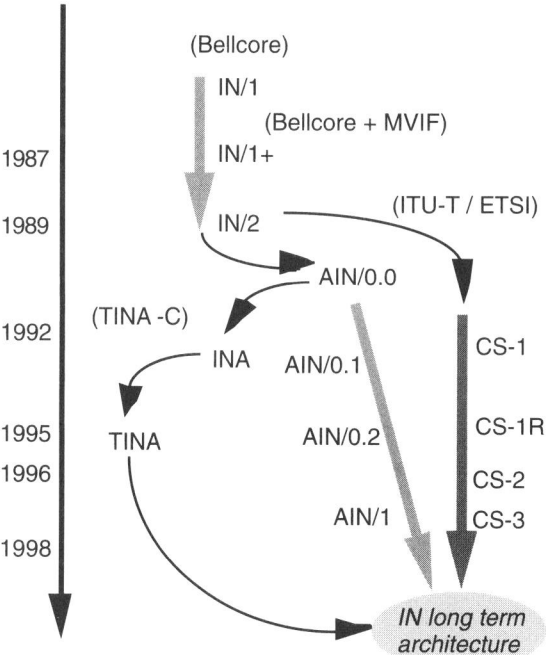

**Figure 2.10** Evolution of IN concepts and standards.

definition of standard interfaces to support any future services. IN-2 was designed as a technical step toward the provision of an infrastructure for deregulation, also known as 'open network architecture', for the public switched network in the US. Since migration from IN-1 to IN-2 implied significant changes in the service switching nodes, Bellcore has defined an interim stage between IN-1 and IN-2, called IN-1+ (Bellcore, 1988), which should establish the initial platform required for full achievement of the service-independent concepts embodied in IN-2.

In 1989 Bellcore has proposed a new IN architecture for the 1992–98 time scale, known as *advanced intelligent network (AIN)* (Bellcore, 1994a–c), as a replacement of IN-1+ and IN-2 since a smooth evolution was not possible with the existing network equipment. AIN envisions a short-term to medium-term IN implementation and should evolve by means of a series of releases, which will be developed in cooperation with the Multi-Vendor Interaction Forum (MVIF), which was formed in the US at the end of 1987. On the way toward AIN release 1 (AIN-1), Bellcore has issued two interim releases, AIN 0.1 (Bellcore, 1992a,b) and AIN 0.2 (Bellcore, 1993a,b). AIN 0.1 was issued in 1992, targeted for implementation in 1993–95. AIN 0.2, which should continue the progress of release 0.1 and carry AIN even closer to its objectives, is targeted for implementation in 1995–97. Additional releases, such as AIN-2, are planned for the long term. However, depending on the progress in the standards bodies, future releases may be replaced by international standards or other long-term-oriented IN architectures, such as Bellcore's information networking architecture (INA) or the telecommunication information networking architecture (TINA), described below. In this book we will address AIN briefly in section 3.4, since the main focus of this book will be on the international IN standards as described below.

In parallel with Bellcore's AIN activities, the international standardization of INs started in 1989 within the International Telecommunications Union (ITU; formerly known as Committ[ac]e Consultatif Internationale de Telegraphique et Telephonique, CCITT) Study Groups XI and XVIII and the European Telecommunications Standardization Institute's (ETSI) Network Aspects Group 6 (NA6). Both bodies aim for the definition of an intelligent network *long-term architecture* (LTA) in stages and focus their work on the development of recommendations for a series of upwardly compatible IN capability sets (CSs). Each capability set is defined in terms of the services to be supported and the functional architecture supporting these services. The first capability set (CS-1) was approved by ITU and ETSI in 1992 (ITU Recommendation Q.121$x$; ETSI TCR-TR NA-60204) and a revised version (CS-1R) was published in spring 1995.

Future capability sets, such as CS-2 and CS-3, should be finalized in 1997 and 1998 respectively. However, the targeted dates for CS-2 and CS-3 may change slightly in the course of standardization. In Chapter 3 we will look in more detail at the first two capability sets.

Besides the international standards bodies, other organizations have investigated the definition of a long-term IN architecture, taking into account the impacts of emerging international *telecommunications management network* (TMN) (ITU Recommendation M.3010) and *open distributed processing* (ODP) (ITU Recommendatio X.900) standards. In general, two work programs have to be recognized in this context: Bellcore's *information networking architecture* (INA) (Bellcore, 1993c) and the international *telecommunication information networking architecture* (TINA) consortium (TINA, 1993). More information on this subject will be provided in Chapter 5.

## 2.5 Toward a generic IN framework – the INCM

In general, the IN can be considered as an additional (network) layer on top of any bearer network, such as a public switched telephone network (PSTN), an integrated services digital network (ISDN), or a broadband ISDN (B-ISDN). In this respect the IN provides a service-oriented network architecture that separates in principal service control functions from service switching functions, with typically both types of functions being implemented in different physical equipment. This is supported by a clear definition of the relationships between these functions, thus providing for network and vendor independence. Because of this separation of functions it is possible to introduce new services rapidly without having to change the functionality of the switches. This means that the IN architecture allows operators to deploy and provide new services more quickly, which is essential in a liberalizing market of ever-increasing competition.

However, another major target of IN is service independence. During the development of many advanced telecommunication services it became clear that all of these services contain similar functionality, i.e. are based on a set of 'service components'. Hence, the idea was to identify a generic sets of reusable service components that could be (re)used for the construction of a new service. Examples of such service components are *authentication, screen, user interaction, number translation, charge*, etc. In this context there is often the notion of an IN programming interface that can be used for easy service creation. What is meant is that a service designer can make use of these service components, by combining them in order to 'implement' a new service. These service

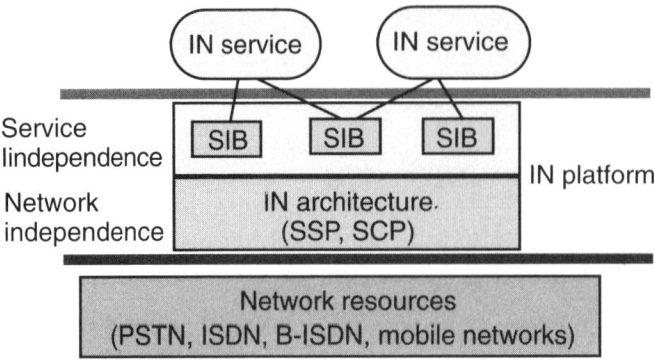

**Figure 2.11** IN platform providing network independence and service independence.

components are referred to as *service-independent building blocks* (SIBs). The resulting 'service (logic) program' could be loaded into the network, i.e. into one or more SCPs, and the new service is instantly available.

Putting these aspects together it can be recognized that the IN is more than just a new network architecture, as the IN aims to support both service and network independence, as illustrated in Figure 2.11.

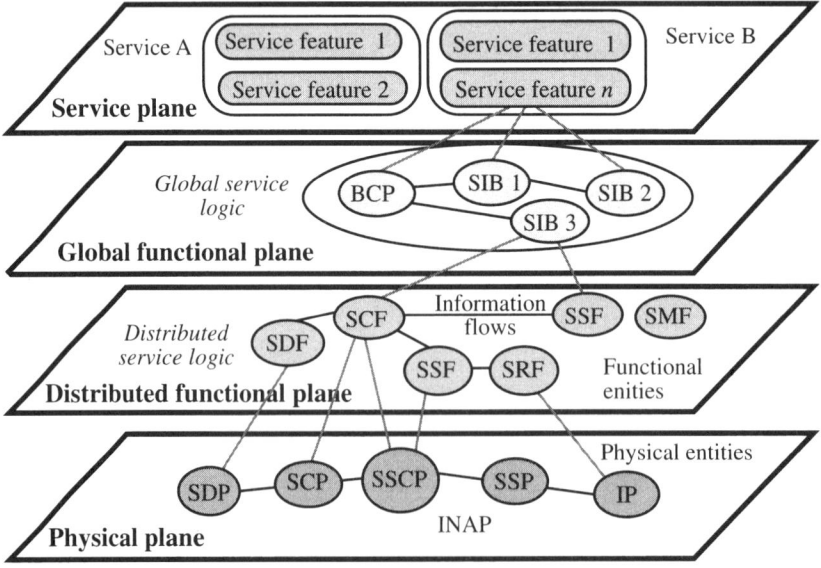

**Figure 2.12** IN conceptual model.

It was felt that the whole engineering process of IN should be captured in a reference model known as the *IN conceptual model* (INCM). This model is the basis for the development of IN standards in ITU (ITU Recommendation Q.1201) and ETSI (TCR-TR NA-60106), which are covered in detail in the next chapter (note that AIN specifications are not based on the INCM). The basic idea of this model is to define a 'top-down' approach for the definition of IN architectures based on IN service capabilities to be supported. Thus the INCM defines four planes addressing service design aspects, global and distributed service provisioning functionality, and physical aspects of an IN-structured network.

It must be stressed that the INCM, as illustrated in Figure 12.2, is only a modeling tool for describing the capabilities and characteristics of an IN-structured network and is not an IN architecture in itself. The reason for this is that only the lower two planes of the INCM address the IN architecture, whereas the higher two planes focus on the creation and implementation of IN services, independent of any (IN) architecture.

Thus, the planes of the model should be regarded as a 'road map' to be followed when defining an IN architecture. Following the model, one has to start with the definition of the services that should be supported in a network-independent way. Basically this is done by identifying the services' basic capabilities (called service features), which can be combined in order to build the identified and also other services. After definition of the functional scope of the service, the INCM leads to a further decomposition of these service capabilities into smaller functional blocks (SIBs) in order to achieve some degree of service independence. This means that the services guide the definition of SIBs, which define basic service capabilities in a network-independent way.

Following the INCM further, the identified basic service capabilities will be used for the definition of an IN architecture. This means that the IN architecture must be able to support the distributed implementation of these service capabilities. The IN architecture definition is subdivided into two stages. In the first stage the network entities are defined in terms of functional elements and their interactions. In the second stage these elements are allocated to specific physical entities. This means that the services can be regarded as the requirements for the architecture definition. Thus, the terms 'INCM' and 'IN architecture' must be distinguished, as INCM is used to define a particular IN architecture. This is illustrated in Figure 2.12 by looking at the specific scope of each plane in more detail.

The *service plane* (SP) is the uppermost plane and describes services from a user's perspective; the service implementation and the underlying network technology are transparent to the users. Each service consists of one or more generic blocks, called *service features* (SFs), each of which could be a complete service or part of a whole service (ITU Recommendation Q.1202).

The *global functional plane* (GFP) models the network from a global high-level perspective as a single programmable entity, hiding the complexity of the distribution of functions. This plane deals with the service creation and contains SIBs that will be used as standard reusable network-wide capabilities for efficient and fast service feature implementation. A *global service logic* (GSL) describes how SIBs are chained together in order to achieve service features, and describes the interactions between a dedicated SIB, known as *basic call process* (BCP), representing a normal call process from which IN services are launched, and appropriate SIB chains (ITU Recommendation Q.1203).

The *distributed functional plane* (DFP) models the distributed view of an IN in terms of units of network functionality, known as a set of *functional entities* (FEs), such as the *service switching function* (SSF) and the *service control function* (SCF). The DFP provides transparency to the physical network elements. Each SIB of the GFP is decomposed in the DFP into a set of client–server relationships between one or more functional entities in the DFP. Each functional entity may perform a variety of *functional entity actions* (FEAs), which means that SIBs are implemented by a sequence of particular FEAs performed by specific FEs. Some of these FEAs result in *information flows* (IFs), i.e. message exchanges, in cases where different functional entities have to cooperate for the provision of SIBs (ITU Recommendation Q.1204). The basis for IN service execution in the DFP is a generic IN *basic call model*, which provides a tool to model an IN service (call) and to describe the distribution of functions between FEs and FE relationships.

The *physical plane* (PP) models the physical aspects of an IN-structured network and contains the real view of the physical network. The PP defines different *physical entities* (PEs) and their interfaces, where the relevant FEs from the DFP are located (ITU Recommendation Q.1205). A complete FE must be implemented within one single physical entity. However, different FEs can be implemented in the same or different physical entities according to different characteristics of the underlying network technologies (e.g. penetration of CSS7 network) and service-specific access requirements. Examples of physical elements are SSPs and SCPs. The information flows in the DFP are implemented in the PP through a standardized OSI-based application layer protocol; the *IN application protocol* (INAP) (ITU Recommendation Q.1218; ETSI ETS 300374-1). This means that physical entities communicate by means of INAP for IN service execution.

In summary, the INCM can be conceptually divided into two parts. The upper two planes of the INCM focus on service creation and implementation by means of generic service building blocks and provide the desired service independence. The lower two planes define a generic IN service provisioning architecture providing for independence of specific bearer networks, such as PSTN, ISDN, etc.

In the next chapter we will look in detail at the different planes of the INCM and explain their interrelationships by addressing the current IN standards.

# 3 Standards for intelligent networks

The international standardization of INs started in 1989 in both the International Standardization Union (ITU) Study Groups XI and XVIII and the European Telecommunications Standardization Institute (ETSI) Group for Network Aspects (NA6). Both bodies aim to develop recommendations standards for a series of upwardly compatible IN capability sets (CSs). The ITU documents are referred to as the 'Q.1200 Recommendation Series for Intelligent Network Architectures' (ITU Recommendation Q.1200). (The ITU recommendations can be obtained directly from the ITU Sales Section, Place de Nations, CH-1211 Geneva 20, Switzerland, Tel +41 22 730 5111, Fax +41 22 730 5194.) In parallel, ETSI defines European technical standards and reports that are consistent with these recommendations but do not directly adopt the ITU Q-series recommendation structure (ETSI TCR-TR NA-60106). (ETSI standards can be obtained directly from the ETSI Secretariat, F-06921 Sophia Antipolis Cedex, France, Tel +33 92 94 4200, Fax +33 93 65 4716.)

The term capability set refers to the set of services and service features that can be constructed by using the SIBs contained in a particular evolution phase of the IN. In March 1992 the first capability set (CS-1) was approved (ITU Recommendation Q.121x; ETSI TCR-TR NA-60204), however a revised version of the first set, sometimes referred to as CS-1R, was released in May 1995 (ETSI TCR-TR NA-6050x). Work on CS-2, addressing basic aspects that were excluded from CS-1, such as IN interworking and management, was started in 1994 and

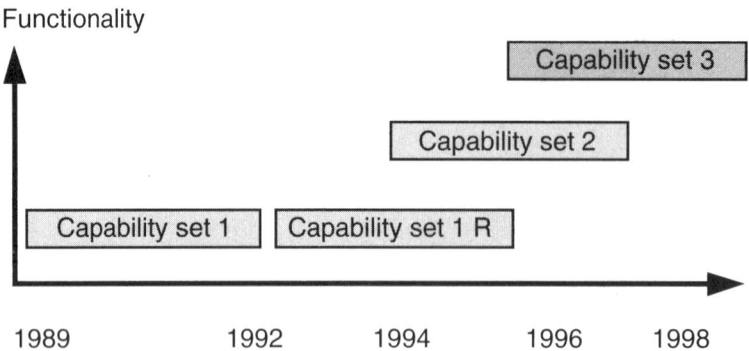

**Figure 3.1** IN standards development schedule (ITU/ETSI).

is scheduled to be finalized in spring 1997. Furthermore, work on CS-3 also started in 1995 and should be finalized in winter 1998. Figure 3.1 illustrates the schedule for the IN capability sets. However, taking the experiences gained from CS-1 standardization into account, it should be stressed that this timing is very ambitious. Thus, it seems likely that CS-2 and CS-3 finalization may be later than targeted.

As outlined in the previous chapter, the INCM is the general framework for the development of international standards in the field of IN. This means that for each capability set all planes of the INCM have to be considered. It is an enhancement of the ITU Recommendation I.130, which defines a 'function-oriented' three-stage methodology for the definition of ISDN services and network capabilities. Consequently, the INCM is the modeling technique for IN, i.e. it describes how to define and realize IN services and the corresponding network capabilities for each capability set. Hence, for each capability set the following stages of work can be identified.

- The stage 1 methodology is used to define services and service features in the service plane and the SIBs in the GFP.
- The stage 2 methodology addresses the definition of functional network elements and their interactions, required for the distributed implementation of the above-defined SIBs in the DFP.
- Finally, the stage 3 methodology defines the corresponding protocols supporting the above-identified interactions between physical network elements in the PP.

Table 3.1 ITU recommendations for the definition of IN capability sets

| Document no. | Title |
| --- | --- |
| Q.1200 | *General Series Intelligent Networks Recommendations Structure* |
| Q.1201/I.312 | *Principles of Intelligent Network Architecture* |
| Q.1202/I.328 | *Intelligent Network Service Plane Architecture* |
| Q.120/I.329 | *Intelligent Network Global Functional Plane Architecture* |
| Q.1204 | *Intelligent Network Distributed Functional Plane Architecture* |
| Q.1205 | *Intelligent Network Physical Plane Architecture* |
| Q.1208 | *General Aspects of the IN Application Protocol* |
| Q.1290 | *Glossary of Terms used in the Definition of Intelligent Networks* |

For details of this methodology readers are referred to the IN User's Guide for CS-1 (ETSI ETR NA-61010; ITU Recommendation Q.1219). Regarding the terminology used with IN, readers are referred to ITU Recommendation Q.1290.

The definition of the IN conceptual model is given in the initial set of ITU 1200 recommendations, one of which defines the overall INCM concept whereas others introduce the architecture of each INCM plane. These recommendations will be revised with each new capability set definition in order to be state of the art. The recommendations in Table 3.1 represent the foundation for the definition of IN capability sets. (For ETSI standards the foundation is NA-60106, which is based on these ITU documents.)

Based on the IN conceptual model and the related recommendations structure, ITU is defining a set of IN recommendations for each IN capability set, i.e. each capability set is structured in the same way (Figure 3.2). Thus, the documents in Table 3.2 will be available in principle for each capability set. Note that ETSI has adopted a different document structure (see ETSI ETR NA-61010).

In the following sections we look at the concepts defined by these standards in more detail following the INCM planes. In section 3.1, representing the main

Table 3.2 Recommendations for each IN capability set

| Document no. | Title |
| --- | --- |
| Q.12x1 | *Introduction to Intelligent Network Capability Set x* |
| Q.12x2 | *Intelligent Network Service Plane Architecture for CS x (not available for CS-1!)* |
| Q.12x3 | *Intelligent Network Global Functional Plane Architecture for CS x* |
| Q.12x4 | *Intelligent Network Distributed Functional Plane Architecture for CS x* |
| Q.12x5 | *Intelligent Network Physical Plane Architecture for CS x* |
| Q.12x8 | *IN Interface Recommendations for CS x* |
| Q.12x9 | *Intelligent Network Users Guide for CS x* |

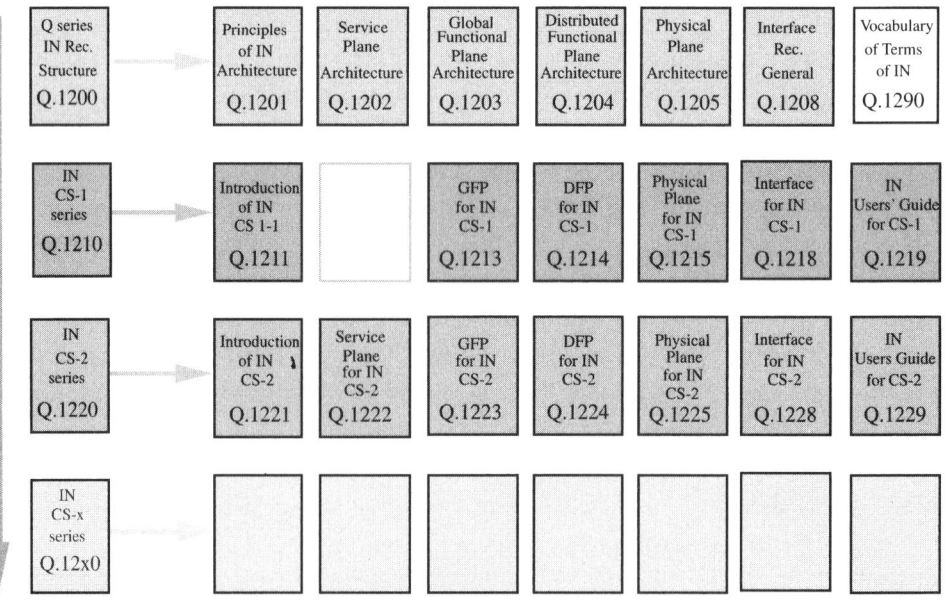

**Figure 3.2** Overview of the ITU IN recommendations series.

part of this chapter, we investigate IN capability set 1 (CS-1) in detail, as this set represents the common ground for IN implementation today. In fact, our considerations are mostly based on the revised version of CS-1, i.e. CS-1R. Note that we do not differentiate between CS-1 and CS-1R unless necessary, i.e. we use the term 'CS-1' even for the revised version.

In section 3.2 we provide some basic information on IN capability set 2 (CS-2) as far as it is available at the time of writing. This information is given in order to illustrate the evolution of the IN. However, it should be stressed that the information may change slightly in the future, as CS-2 standards are still under revision.

In section 3.3 we look briefly at the focus of future capability sets. Section 3.4 provides some information on the North American view of IN standards, represented by Bellcore's advanced intelligent network (AIN) specifications. Finally, we present some information on IN deployment and products in section 3.5.

## 3.1 IN capability set 1

CS-1 (and its revised version, CS-1R) represents the first standardized stage of IN for supporting a first range of IN services. The ITU IN CS-1 recommendations in Table 3.3 are available.

Correspondingly, ETSI has defined a set of IN CS-1 technical reports and standards. Table 3.4 indicates the relationships to the corresponding ITU documents.

Table 3.3 ITU IN CS-1 recommendations

| Document no. | Title |
| --- | --- |
| Q.1211 | *Introduction to Intelligent Network Capability Set 1* |
| Q.1213 | *Global Functional Plane for Intelligent Network CS-1* |
| Q.1214 | *Distributed Functional Plane for Intelligent Network CS-1* |
| Q.1215 | *Physical Plane for Intelligent Network CS-1* |
| Q.1218 | *Intelligent Network Interface Recommendations for CS-1* |
| Q.1219 | *Intelligent Network Users Guide for Capability Set 1* |

Table 3.4 Relationship between ETSI and ITU recommendations

| ITU document | Related ETSI document | Title |
|---|---|---|
| Q.1201 | ETSI TCR-TR NA-60106 | *Intelligent Network: Framework* |
| Q.1211 | ETSI TCR-TR NA-60204 | *Guidelines for CS-1 Standards* |
| Q.1213 | ETSI TCR-TR NA-60501 | *Global Functional Plane for IN CS-1* |
| Q.1214 | ETSI TCR-TR NA-60502 | *Distributed Functional Plane for IN CS-1* |
| Q.1215 | ETSI ETS 300348 | *IN CS-1 Physical Plane* |
| Q.1218 | ETSI ETS 300374-1 | *Core INAP* |
| Q.1219 | ETSI ETR NA-61010 | *IN User's Guide for CS-1* |

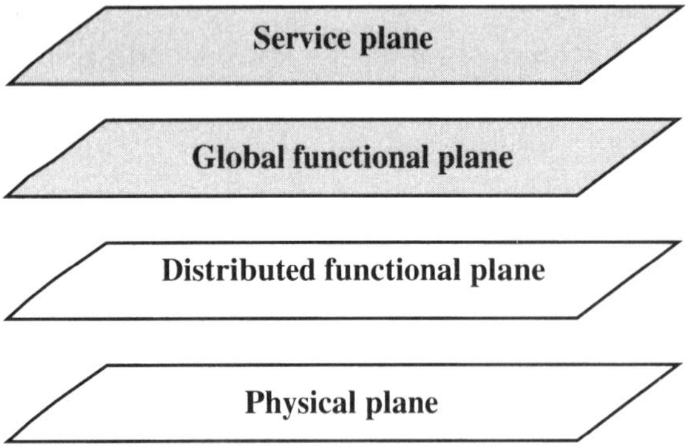

**Figure 3.3** CS-1 service capabilities are defined by the upper two INCM planes.

In the next section we present an overview of the standardized concepts related to CS-1 services and architecture and also look at the upper and lower planes of the INCM.

## 3.1.1 CS-1 services

This section is devoted to the upper two planes of the INCM (Figure 3.3), focusing on the service plane and the global functional plane. We will look in section 3.1.1.1 at the services targeted for CS-1 and the corresponding set of CS-1 service features defined in the service plane. We will also briefly address service (feature) interactions. In addition, we provide in section 3.1.1.2 an overview of how the service features are achieved in the GFP by SIBs.

It must be emphasized that the standardization of 'IN services' is not within the scope of IN standards, as the IN aims for the definition of a generic service execution platform for an 'open' set of services. However, some services, sometimes called 'benchmark' services, have been identified in order to guide the development of IN network capabilities, in other words the identified IN services should be seen as high-level requirements (or the starting point for the stage 3 methodology) for the development of IN capabilities.

### 3.1.1.1 CS-1 service plane

Although, by nature, the IN is a service-independent architecture, the identification of targeted telecommunications services and service features is fundamental to the definition of SIBs, call-processing models and service control principles. The focus of current IN standards and platforms is on supporting telecommunication services. A telecommunication service in the IN context is defined as a service offered to end users by a service provider who is responsible for the service logic. A service constitutes a stand-alone commercial offering, characterized by one or more *core* service features that can be optionally enhanced by other service features. Note that within IN there is no distinction between basic services and supplementary services as seen in other networks, such as ISDN or GSM.

*Services*

IN services are categorized into two types.

1. The first category are *single-ended* and *single-point-of-control* services, referred to as 'type A' services.
2. All other services are referred to as 'type B' services.

'Single-ended' means that the service applies to one and only one party in a call and is independent of any other parties that may be participating in a call. This allows other single-ended services to apply to another party in the same call as long as the service instances do not have feature interaction problems (see below). For example a calling user may use an 'abbreviated dialing' service, whereas the called party may use a 'call-forwarding' service. Both services could be used within the same call without any interference, with one service logic active in the originating call segment and one service logic active in the terminating call segment.

'Single point of control' describes a control relationship in which a call is influenced by one and only one service logic program at any point in time. This means that for type A services there are no interactions between SCPs for the provision of an IN service. Hence, these services are characterized by a relatively simple control relationship between the SSP and the SCP, in which the SSP is the 'client' for service-related information provided by the SCP and the switch retains connection control at all times (see section 3.1.2 for details).

In type B services, in contrast, several IN subscribers may be associated within a single call, and during a call several call parties may be added or dropped dynamically, requiring call topology manipulations. Hence, an SCP has to cope with multiple parties involved in a single call. Also, interactions between different SCPs may occur, as each call party may have its own set of subscribed services. Multiparty services could be regarded as examples of type B services.

CS-1 is targeted to support type A services only to limit the operational complexity! Looking at the services defined in CS-1 one may identify the following categories (note that one IN service may be grouped into several categories):

- number translation services;
- alternate billing services;
- screening services; and
- other services.

Notice that the notion of CS-1 service categories is not found in the standards. These categories have been introduced by the authors for illustration purposes only.

1. *Number translation services.* The primary advantage of this category of services is its flexible numbering and routing capabilities. In addition,

customer control capabilities may be incorporated. The services usually support the called party, but some services are also intended for the calling party.

- *Abbreviated dialing* enables the use of short numbers for outgoing calls.
- *Call forwarding* enables forwarding of incoming calls to another destination.
- *Call rerouting distribution* allows incoming calls to be rerouted in the case of a busy line.
- *Call volume distribution* allows routing of incoming calls to different locations.
- *Destination call routing* allows routing of calls depending on time, origin, caller, etc.
- *Freephone* allows reverse charging for a unique number.
- *Follow-me diversion* allows redirection of incoming calls by remote control.
- *Premium rate* allows information service providers to obtain revenues from calls.
- *Selective call forwarding on busy/don't answer* allows preselected callers to be rerouted in case of a busy line.
- *Universal access number* supports one number for several terminating lines.
- *Universal personal telecommunication* enables a unique number for incoming and outgoing calls at any terminal.
- *User-defined routing* allows personal routing schemes for outgoing calls.

2. *Alternate billing services.* These services take advantage of the flexible charging capabilities. Nevertheless, these services may incorporate some number translation capabilities, but the service focus is on special charging or billing.

- *Account card calling* allows calls from any telephone by charging a specified account.
- *Automatic alternative* billing allows calls from any telephone by charging a separate account.
- *Credit card calling* allows calls from any telephone by charging a credit card.
- *Split charging* allows charge splitting between parties of a call.

- *Premium rate* is described above.

3. *Screening services.* These services take advantage of the flexible screening capabilities in order to restrict call establishment. Nevertheless, they may incorporate some number translation capabilities, but the service focus is on screening capabilities.

   - *Originating call screening* supports restriction of incoming calls.
   - *Security screening* enables screening of users seeking network access.
   - *Terminating call screening* supports restriction of outgoing calls.

4. *Other services.* These are services that cannot be included in any of the other categories, as they provide complementary capabilities. This means that some services also incorporate flexible routing, charging and/or screening capabilities, but the basic service provided is focusing on something particular. In addition, they may incorporate advanced information handling capabilities.

   - *Completion of call to busy subscribers* supports automatic call back when the line becomes free.
   - *Conference calling* allows multiple parties within a single call.

   These two services are only partly supported by CS-1, as they require additional network capabilities.

   - *Malicious call identification* enables logging of incoming calls.
   - *Mass calling* supports handling of high call volume.
   - *Televoting* enables voting via the network.
   - *Virtual private network* simulates a private network by using public network resources.

A more detailed description of all these CS-1 services is given in section A.1. For more details readers are referred to ITU Recommendation Q.1213 and ETSI TCR-TR NA-60501. However, note that these services are only briefly described, as a detailed specification is outside the scope of the standards!

*Service features*

An IN service feature reflects a specific aspect of the functionality of an IN service, i.e. a service feature is a specific part of a telecommunication service that can also be used in conjunction with other telecommunication services/service features as part of a commercial offer. In other words, this means that the identified services features can be arbitrarily combined to define services (as long as it results in a sensible service). This may result in new services that have not been identified in the service plane.

Generally, IN service features can be divided into the following categories (notice that such notion of CS-1 service features categories are not contained in the standards). These categories are introduced by the authors for illustration purposes only.

1. *Numbering features* allow the use of dedicated numbers for making calls.

   - *Abbreviated dialing* allows the use of short numbers for outgoing calls.
   - *One number* allows two or more destination lines to be reached via one number.
   - *Personal number* allows individuals to have a unique universal personal telecommunications (UPT) number for incoming/outgoing calls.
   - *Private numbering plan* allows VPN users to use their own numbering plan.

2. *Routing features* allow the destination number to which an incoming call should be routed to be determined.

   - *Call distribution* allows incoming calls to be routed to one of multiple destinations.
   - *Call forwarding* allows incoming calls to be routed to another destination.
   - *Follow-me diversion* allows registration for incoming calls at any telephone.
   - *Time-dependent routing* allows incoming calls to be routed based on time, day, week, etc..
   - *Origin-dependent routing* allows incoming calls to be routed based on call origin area.

3. *Charging features* permit control service charging.

   - *Premium charging* allows sharing of revenue from extra call charges.
   - *Reverse charging* allows call charges to be allocated to the called party.

- *Split charging* allows the call charges to be split between the calling and called parties.

4. *Access features* supply functions for access control.

   - *Authentication* allows verification of a user.
   - *Authorization code* allows calling restrictions of a terminal to be overridden.
   - *Off-net access* allows VPN users to call from outside into the VPN.
   - *Off-net calling* allows VPN users to call outside the VPN.

5. *Restriction features* support the screening of calls regarding various conditions.

   - *Call limiter* allows a subscriber to limit the number of calls to a destination line.
   - *Call gapping* allows the service provider to restrict incoming calls to a destination line.
   - *Closed user group* allows outgoing and/or incoming calls only within a specific group.
   - *Originating call screening* allows incoming calls to be restricted.
   - *Terminating call screening* allows outgoing calls to be restricted.

6. *Customization features* allow the customer to define and modify service parameters.

   - *Customer profile management* allows subscribers to modify their service data.
   - *Customer recorded announcement* allows the caller to be prompted by a customized voice message.
   - *Customized ringing* allows specific ringing to be assigned to specific calling parties.

7. *User interaction features* support dialogues with a call party during call set-up.

   - *Attendant* allows a VPN user to obtain special help.
   - *Consultation calling* allows another call to be placed during an ongoing call.
   - *Destination user prompter* allows the called party to be prompted.
   - *Originating user prompter* allows the calling party to be prompted.

8.  *Other service features* cannot be allocated to one of the above categories. Most of these features provide specific call-handling capabilities.

- *Automatic call back* allows a call to be set up after a line becomes free.
- *Call hold with announcement* allows a call to be placed on hold with prompting.
- *Call logging* allows call information to destination lines to be kept.
- *Call queuing* allows incoming calls to be queued when a called line is busy.
- *Call transfer* allows a call to be transferred to another destination line.
- *Call waiting* alerts a busy called party to another incoming call.
- *Mass calling* allows processing of a huge number of incoming calls.
- *Meet-me conference* allows resource reservation for multiparty calls.
- *Multiway calling* allows multiple simultaneous calls.

Section A.2 provides a more detailed description of these CS-1 service features. It has to be recognized that these service features are only defined in words (this is referred to as a 'stage 1' description by the standards bodies) in the standards (ITU Recommendation Q.1213; ETSI TCR-TR NA-60501), which allows for varying implementations of each service feature and thus the services themselves.

A service feature is either a *core part* of a telecommunication service, which means that it is fundamental to the telecommunication service, or an *optional part* offered as an enhancement to a telecommunication service. This means that the basic functionality of an IN service is defined by its core service feature(s), i.e. in the absence of the core service feature(s) the name of the IN service does not make sense as a commercial offering.

By adding or dropping (optional) service features, different variations of an IN service can be created. Basically this concept enables the customized/tailored provision of IN services, whereby several variations of one IN service, containing different sets of optional service features, may exist in parallel. Section A.3 provides an overview of which core service features are present in each CS-1 IN service and which optional service features could be added to enhance the basic service functionalities.

For example, the basic Freephone service is defined by the two core features 'one number' and 'reverse charging' which are crucial for that service. In addition, the Freephone service may incorporate many optional features, such as 'call limiting', 'call queuing', and 'time-dependent routing', different combinations of these optional service features allowing for different customized Freephone services to meet specific customer demands. One Freephone service offer may comprise the service features 'one number', 'reverse charging', 'call limiting', and 'call queuing', whereas another may comprise the service features 'one number', 'reverse charging', and 'time-dependent routing'. This example is elaborated further in section 3.1.4.

*Service feature interactions*

The issue of service interactions, also referred to as *'service feature interactions'*, is one major technical problem of IN that is as yet unsolved. The basic problem stems from the implicit assumption of IN that services are designed for a network regarded as a virtual computer, with (service) programs having virtual access to all the computer memory and ignoring the other programs. It is assumed that resources are freely available to a service program and that other service programs can be ignored. However, it is exactly this issue that is the source of the problems!

For example, let us consider the relationships between a 'call-forwarding' service and a 'Freephone' service that are available in the same network. A single service subscriber would like to make use of both services. Possible interactions between both services may be:

- Freephone calls are not forwarded;
- Freephone calls are forwarded and the first called party has to pay;
- Freephone calls are forwarded and the forwarded-to party has to pay, etc.

The only solution to this problem is for the service designer, when a new service is introduced, to define its interactions with all other existing services. For example, if 'call forwarding' is already implemented in the network, the introduction of a 'Freephone' service requires a clear definition of possible interactions between Freephone and call forwarding. In addition, the Freephone service designer must define the interactions with all other services, such as 'card calling' or 'call waiting'.

Furthermore, service interaction may occur with more than one service at a time: one call may involve several services or features, and different service combinations may involve different ways of solving interactions. For example, the service interactions involved when service B is invoked after service A, and service C after service B may be different from those that occur when service C is invoked after service A, and service B after service C.

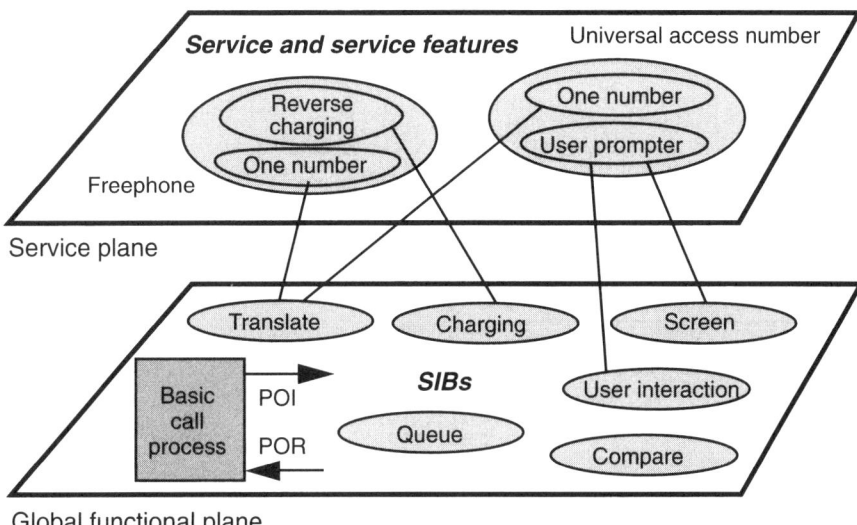

**Figure 3.4** SIBs realizing service features.

These examples indicate that service interactions must sometimes be defined at the service level, which makes it much more difficult to specify services. Up until now, service interaction has been solved at the service specification phase, i.e. the specification of a service usually also includes the procedures for solving the interactions with other already specified services. However, this approach only works if the number of services is limited. Thus, one important problem to be solved by the IN is the ability to support generic mechanisms for solving these interaction problems. Current IN research is addressing this problem (IEEE, 1993a; Feature Interactions in Telecommunications Software Systems, 1995) and CS-2/3 standards will tackle this problem in more detail.

### 3.1.1.2 CS-1 global functional plane

CS-1 defines a high-level logical programming interface designed for use by the service designer by means of a *service creation environment* for the definition of *service logic programs* (SLPs). This logical programming interface is composed of a set of SIBs in the GFP (ITU Recommendation Q.1213; ETS TCR-TR NA-60501). SIBs are standard reusable network-wide capabilities (e.g. translating numbers, collecting digits from a user, perform screening, etc.) used to create IN services, i.e. service features. They provide significant flexibility for rearranging defined network functionality to create new service features and to customize existing service features. This means that a service feature is described in terms of one or more SIBs as illustrated in Figure 3.4.

Section A.3 provides an indication of which SIBs may be involved in the implementation of the service features identified in the service plane. However, it must be stressed that no exact mapping of service features onto SIBs exists, as there is no exact specification of IN service features, only a stage 1 description. Thus, the specification and the resulting implementation of a particular service feature may vary, which means that the use of SIBs for service feature implementation may also vary.

An SIB has the following characteristics.

- It is a reusable building block, describing a single complete activity.
- It has a unified and stable interface, with one or more inputs and one or more outputs.
- It is completely independent of any physical architectural considerations, i.e. the network is viewed globally as a single virtual machine.

The CS-1 standards define the following SIBs.

- *Algorithm:* applies a mathematical algorithm to data in order to produce a result.
- *Authenticate:* provides authentication functionality for a service.
- *Charge:* is used to determine special charging treatment for the call when modified charging is required (i.e. reverse charging, split charging, etc.).
- *Compare:* performs a comparison of an identifier against a list of specified values. Three results are possible – greater than, less than, or equal to.
- *Distribution:* allows the distribution of calls to different logical ends of the SIB according to user-specified parameters.
- *Limit:* limits the number of calls that are allowed through an IN-structured network by filtering IN-related calls with given characteristics, even although they are not causing congestion. This SIB may be used to block all or a fraction of calls. Such limitations are based on service provider-specific parameters.
- *Log call information:* records detailed information; information may be collected for management purposes, but not for call-related purposes (i.e. routing).
- *Queue:* provides the sequencing of calls to be completed to a called party. This SIB places a call that cannot be completed in a queue,

which involves queuing incoming calls to a given party, dequeuing calls, and timing the calls in the queue for possible alternative treatments.

- *Screen:* performs a comparison of an identifier against a list to determine whether an identifier has been found in the active list. This SIB can be used for authorizing the completion of a call for the calling party or called party, originating or terminating call screening, etc.
- *Service data management:* provides the ability to modify, store, and retrieve information associated with a service subscriber. This SIB includes capabilities related to all the aspects of information resource management within the network, including actions such as add, retrieve, and delete information.
- *Status notification:* provides the ability to inquire about the status, or changes in the status, of network resources. There are three types of notification status requests – polling resource status, monitoring for changes, and continuous monitoring.
- *Translate:* determines an output information from some input information and additional features, based on various other input parameters. For example, a dialed number plus time of day and calling party identification may be translated to a (physical) destination number.
- *User interaction:* allows the exchange of information between a call party (calling or called party) and the network.
- *Verify:* provides confirmation that the information received is syntactically consistent with the expected form of such information. This SIB can be used in user interaction and network addresses.
- *Basic call process:* a dedicated SIB responsible for providing basic call connectivity between parties in the network. This SIB provides the basis for flexible call-handling capabilities and will be described below in more detail.

It has to be noted that ETSI has defined additional SIBs within CS-1 to the ITU SIBs (ETSI TCR-TR NA-60501, ETSI ETR-TR NA-61010):

- *Connect* completes a call to a defined destination.
- *Continue* continues basic call processing from the point at which it was suspended.
- *Disconnect resource* releases all specialized resources.
- *EDP info* performs the retrieval of an event detection point (EDP).
- *EDP request* arms EDPs in the basic call process.

- *Initiate call* initiates a call to a defined destination.
- *Release call* releases a call during any phase of call processing.

In general, SIBs are independent of any service feature in which they are used. However, some elements of service dependence are needed in order to achieve the SIB functionality within a specific service feature. Thus, data parameters are used to tailor SIBs to perform a desired functionality. These data parameters are made available to the SIB through global service logic (described below).

Two types of data parameters are required for each SIB, as depicted in Figure 3.5.

- Dynamic parameters, called *call instance data* (CID), whose value will change with each call instance, will be used to specify call- and/or subscriber-specific details, e.g. calling or called line information.
- Static parameters, called *service support data* (SSD), are specific to the service feature that the SIB is enabling. Besides fixed parameters, whose values are fixed for all call instances, the SSD contains a specific pointer, identifying which CID is required by the SIB.

Examples of CID are call reference, calling line identity, calling line category, dialed number, destination number, bearer capability, etc.

In general, an SIB description contains the following information:

- a text definition of the SIB (from a service creation point of view);
- a description of the actions performed by the SIB (i.e. operations);
- an identification of potential services the SIB can be used for;

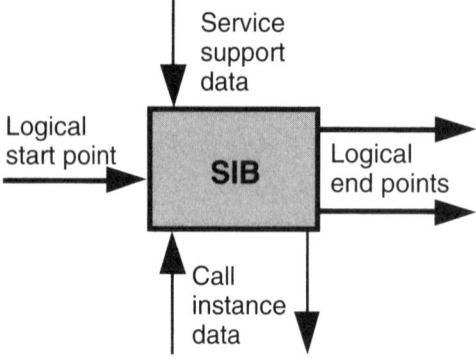

**Figure 3.5** Graphical representation of a SIB.

- the required input, i.e. CID, SDD, and a logical start point;
- the output, i.e. CID and one or more logical end points (e.g. success or error);
- a graphic representation; and
- a specification description language (SDL) diagram.

It is assumed that SIBs are combined by chaining them, thereby implying an algorithmic approach. IN service logic built with SIBs, i.e. an SIB chain, is referred to as *global service logic* (GSL). The GSL represents the 'glue' that describes the order in which SIBs can be chained together to accomplish service features. This SIB chain has to interact with the basic call at specific points in call processing.

Thus, a dedicated SIB is used to model the network real-time behavior and provides the basis for flexible call-handling capabilities, known as the *basic call process* (BCP). It is the BCP that allows the passing of call control temporarily to IN service logic, i.e. to an external SIB chain. This means that the BCP SIB provides IN service logic with elementary call-processing capabilities. For CS-1 these capabilities include:

- *call set-up* capabilities to influence originating or terminating call set-up for two-party calls (e.g. for flexible routing, call queuing, call diversion);
- *call party handling* capabilities enabling the handling of individual call parties (e.g. hold/retrieve parties in a call or add/drop parties from a call);
- *call initiation* capabilities to initiate calls between two parties;
- *call clearing* capabilities to release calls; and
- *event reporting* capabilities to request the reporting of call-processing events (e.g. caller abandon, busy or no answer).

To provide these capabilities to IN service logic, or in other words to allow the enhancement of these basic call capabilities by means of IN service logic, the BCP SIB provides specific interaction points. *Points of initiation* (POI) allow the transfer of control from the BCP to the GSL , whereas *points of return* (POR) allow the transfer of control back to the BCP. Within the GFP there is no spatial distribution of activities. However, projecting this concept to the DFP described in the next section, it has to be recognized that the BCP runs on a switch and the GSL runs on an SCP. Hence, a POI marks a point at which control is transferred from the switch hosting the BCP to an SCP hosting the GSL, whereas a POR represents a point in service logic at which control is returned from the SCP back to the switch.

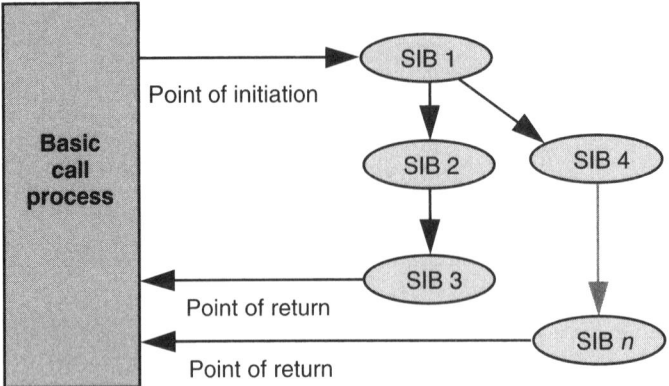

**Figure 3.6** Global functional plane model.

One may compare this approach of launching external service logic to a remote procedure call, but it is quite different, as there may be several POR for one POI, and there may be additional interactions between the BCP and GSL between one POI and one POR (e.g. events). This is indicated in Figure 3.6.

The identified POI within CS-1 are as follows.

- *Call originated* identifies that the user has made a service request without yet specifying a destination address (e.g. off-hook but before dialing).
- *Address collected* identifies that the address (i.e. the dialed number) input has been received from the user.
- *Address analyzed* identifies that the address input has been analyzed to determine characteristics of the address (e.g. a Freephone number).
- *Call arrival* identifies that the network is prepared to attempt completion of the call to the terminating party.
- *Busy* identifies that the call is destined to a user who is currently busy.
- *No answer* identifies that the call has been offered to a user who has not answered.
- *Call acceptance* identifies that the call is active but the connection between calling and called parties is not established (i.e. called party off-hook but no switch-through).
- *Active state* identifies that the call is active and the connection between the calling and called parties is established.
- End of call identifies that a call party has disconnected.

The identified POR within CS-1 are as follows.

- *Continue with existing data* identifies that the BCP should continue call processing with no modification.
- Proceed with new data identifies that the BCP should continue call processing with a data modification.
- *Handle as transit* identifies that the BCP should treat the call as it has just arrived.
- *Clear call* identifies that the BCP should clear the call.
- *Provide call party handling* identifies that the BCP should perform functions to enable call control for individual call parties (this POR requires further study).
- *Initiate call* identifies that the call should be initiated.

Putting the pieces together, for a given service feature the GSL describes:

- a specific POI that will define the functional launching point from the BCP to the SIB chain;
- a specific set of POR in which the SIB chain can logically return to the BCP;
- the pattern and order of SIBs that are to be chained together (beginning with a POI and ending with a set of POR); and
- data parameters (i.e. SSD and CID) for each SIB in the SIB chain.

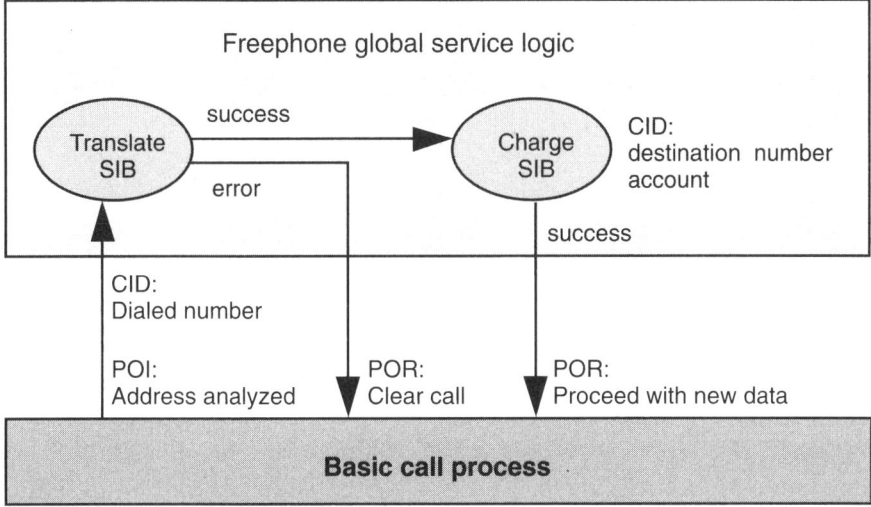

**Figure 3.7**  Example GSL for a Freephone service.

For a simple Freephone service (see Figure 3.7) containing only the core features 'one number' and 'reverse charging', the GSL contains the chained SIBs 'translate' and 'charge' starting at the POI 'address analyzed' and returning at the POR 'proceed with new data' in the case of success to the BCP. In the case of an error, the GSL returns at the POR 'clear call' to the BCP. The used CID contains, among other things, the dialed number, subscriber account, and (physical) destination number.

Once the GSL is defined in the GFP, it must be transformed into *distributed service logic* (DSL) in the DFP to take into account distribution aspects in the IN functional architecture. However, DFP interactions are transparent to the SIBs in the GFP.

In summary, SIBs are considered as the foundation of the 'IN programming interface'. This means that, by combining SIBs into SIB chains, new services, i.e. service features, can be easily and rapidly constructed. In addition, the manipulation of SIB chains allows for the customization of service features. In this context the notion of 'service creation environments' is important as they provide the service designer with an appropriate graphical–user interface. However, new IN services features can only be implemented within the capabilities of the available set of SIBs, as the definition of new SIBs is a complex issue requiring corresponding enhancements within the underlying planes of the INCM.

Some people believe that SIB specifications are not yet at the stage at which multivendor implementations are possible. The reason for this is that some SIBs are considered to be to general (such as the 'algorithm' SIB) or represent too large aggregations of functionality (e.g. the 'translate' SIB). However, work on SIB identification and specification is progressing (see also section 3.2.1.2).

## 3.1.2 CS-1 architecture

The IN architecture is defined by the lower two planes of the INCM, namely the DFP and the PP (Figure 3.8). Thus, the standardized IN architecture is separated into a functional architecture and a physical architecture, both of which will be described in this chapter.

### 3.1.2.1 CS-1 distributed functional plane

The functional IN architecture for CS-1 defined in the IN DFP (ITU Recommendation Q.1214; ETSI TCR-TR NA-60502) is an abstract representation of the architectural functions that support the provision of IN services, defining both the FEs and the relationships between them. In this section we

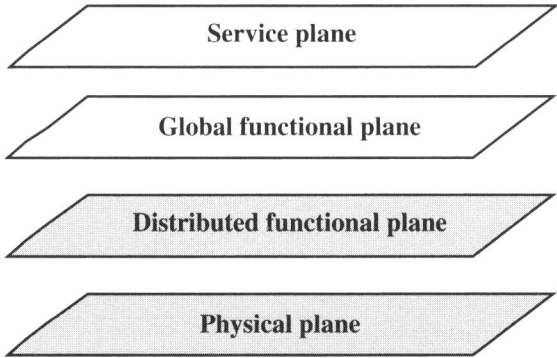

**Figure 3.8** The CS-I IN architecture is defined by the lower two INCM planes.

introduce the basic FEs and describe how the SIBs defined in the GFP are implemented by the entities in the DFP. In addition, we introduce in section 3.1.2.1.3 the basic call model, representing the prerequisite for the 'remote control' of switches and thus for the distributed implementation of IN services.

*Functional entities*

The functions that have been defined as a first subset of a target IN architecture are related to traditional call handling (service switching/triggering), service execution (service control) and service management, as depicted in Figure 3.9.

1. *Basic call-handling functions.* The *call control agent function* (CCAF) represents the user terminal function and hence provides access to the network. The CCAF accesses a *connection control function* (CCF) that provides basic call-processing functionality and can thus be considered as a traditional 'switch'. To perform this basic call processing, the CCF makes use of a 'basic call model', which will be discussed below in more detail. Both the CCAF and the CCF represent the basic elements of existing telecommunications networks, e.g. PSTN or ISDN.

2. *Service execution functions.* To provide supplementary services according to the IN concept, a number of additional functions are needed. A *service switching function* (SSF) represents additional functionality for controlling switch resources and provides a well-defined, service-independent interface to the *service control function* (SCF) that controls

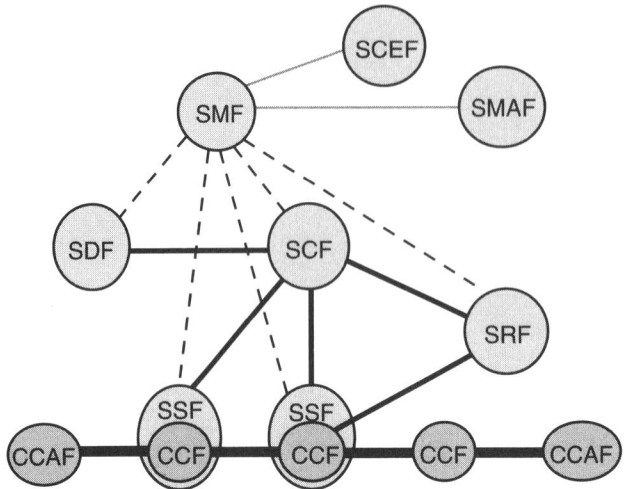

**Figure 3.9** Distributed functional plane model for CS-1.

resources in a switch or peripheral based on an appropriate service logic program. A *service data function* (SDF; sometimes also referred to as specialized data function) contains the service data (customer- and network-related data) and provides standardized real-time access for SCFs to service data. Additional functions for controlling (intelligent) peripheral resources such as tape recorders and speech synthesizers, used for collecting information from subscribers or playing announcements, are provided by a *specialized resource function* (SRF).

3. *Service management functions.* The *system management function* (SMF) supports service introduction, provision, and maintenance and is accessed by a *service management agent function* (SMAF), providing the man–machine interface to the SMF. An additional *service creation environment function* (SCEF) allows for the specification, testing, and introduction of services in the IN. Output of this function involves, among other things, a service logic program (for the SCF), a service data template (for the SDF), and service trigger information (for the SSF). (SMAF, SMF and SCEF are not covered in detail within CS-1.)

The role of each of the basic call-handling and service execution-related functions for supporting IN services is as follows. A CCAF receives call set-up/service requests (e.g. 'off-hook' indicator or dialed digits) from end users and passes them on to a CCF/SSF for processing. The CCF/SSF processes these requests to determine, based on existing conditions and predefined criteria, if they are requests for IN services. If so, for a given request, the SSF passes the request to an SCF, along with the state of the call/service attempt at the time the request was detected. The SCF invokes and processes the appropriate IN service logic program for the requested service based on the state of the call/service attempt, and interacts with the SDF (optionally with an SRF) and the CCF/SSF to provide the requested service to the end user.

Note that CS-1 defines only the SSF–SCF, SCF–SDF, and SCF–SRF interface, with the primary focus on the SSF–SCF interface. The service management control relationships of the model, i.e. the interfaces between the SMF and the other functional entities, are not yet specified.

In summary, a specific IN service in the DFP consists basically of three parts, namely specific service trigger information in the SSF(s), a specific service logic program in the SCF, and service-specific data in the SDF. For some service features making use of the 'user interaction' SIB, there may also be some specialized data, i.e. customized announcements, in the SRF.

*Service modeling in the DFP*

As outlined in section 3.1.1, IN services, i.e. service features, are composed of SIBs in the GFP. The monolithic view of a SIB in the GFP has to be decomposed in the DFP into an interacting set of capabilities. Each functional entity in the DFP may perform specific operations, referred to as *functional entity actions* (FEAs). Thus each SIB is decomposed in the DFP into a set of client–server relationships between one or more functional entities, with the client being the service control function and the server being one of the other functional entities, such as the service data function, specialized resource function, or service switching function.

Consequently, different functional entities in the DFP must exchange messages to perform a desired SIB functionality (Figure 3.10). These client–server information exchanges between the functional entities are called *information flows* (IFs). The total set of IFs between any two functional entities in the DFP will be a number of such client–server information flows.

Within CS-1 about 40 IFs have been identified. Some examples of functional IFs are as follows.

- *Initial detection point:* the SSF starts a dialogue with the SCF and requests further instructions (based on the occurrence of a specific trigger event).
- *Play announcement:* the SCF instructs the SRF to send an announcement to a user.

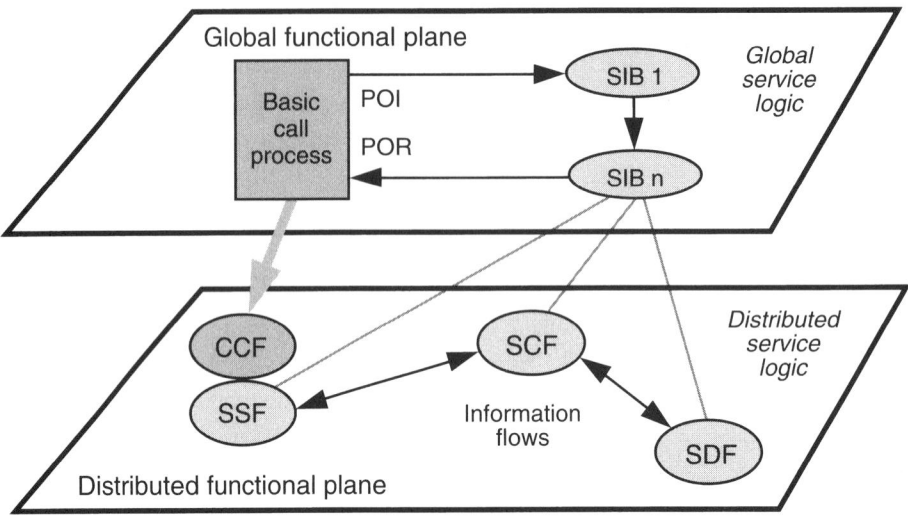

**Figure 3.10**   Realization of SIBs by information flows in the distributed functional plane.

- *Prompt and collect user information:* the SCF instructs the SRF to collect some dialed information from the user with some guiding prompts.
- *Query:* the SCF consults the SDF to translate a logical number to a specific destination number.
- *Connect:* the SCF guides the SSF to connect a call to a specific destination.
- *Apply charging:* the SCF provides specific charging information the SSF has to use, etc.

Table 3.5 indicates which FEs are involved in the implementation of each CS-1 SIB identified in the GFP, i.e. it illustrates which IN functional entities have to interact, i.e. exchange information flows, for the implementation of a particular SIB. It can be recognized that the SCF is involved in all IFs, as it hosts the service logic program, keeping control of all IFs.

For example, the BCP SIB is achieved by a specific set of IFs between the CCF/SSF and the SCF. A comprehensive list of all IFs related to implementation of each SIB in the DFP is given in ITU Recommendation Q.1214 and ETSI TCR-TR NA-60502.

Nevertheless, the starting point for the execution of IN service logic achieved through IFs in the DFP represents a generic 'call model', describing the distribution of functions between functional entities and functional entity

Table 3.5 Functional entities involved in the realization of SIBs

| SIB | Functional entity | | | | |
| --- | --- | --- | --- | --- | --- |
| | Connection control function | Service switching function | Service control function | Specialized resource function | Service data function |
| Authenticate | | | ● | | ● |
| Algorithm | | | ● | | |
| Charge | ● | | ● | | |
| Compare | | | ● | | |
| Distribution | | | ● | | |
| Limit | ● | | ● | | |
| Log call information | ● | | ● | | ● |
| Queue | ● | | ● | ● | |
| Screen | | | ● | | ● |
| Service data management | | | ● | | ● |
| Status notification | ● | | ● | | ● |
| Translate | | | ● | | ● |
| User interaction | ● | | ● | ● | |
| Verify | | | ● | | |
| Basic call process | ● | | ● | | |

relationships, and in particular identifying all possible points in basic call processing from which service control may be passed from a switch to an IN service logic.

*Basic call model*

The primary goal of IN services is to enhance the functionality of basic tele-communications services, such as telephone calls. To achieve this, IN services have to control network resources, i.e. the switching facilities, in a flexible and efficient way. Thus, a standard resource control interface, i.e. the SSF–SCF interface, has been developed. This approach requires the switches, i.e. the connection control function/service switching functions, to be capable of providing visibility and control on detailed call events. Hence, a corresponding model is required within the DFP that identifies all possible points in basic call processing, as seen in a switch, from which IN services can be invoked, i.e. when interactions between the SSF and SCF can take place.

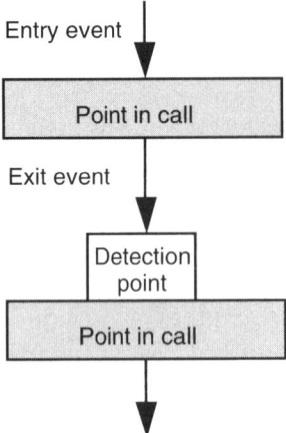

Entry event

Point in call

Exit event

Detection point

Point in call

**Figure 3.11**  Relationship between detection points and points in call in the basic call state model.

This model is called *basic call model* (BCM); it represents a standardized view of call-processing functions to external service logic, and as such provides the framework for IFs between the SSF and SCF . This means that IN service logic located in the SCF 'sees' a call only by means of the information it receives from the SSF, i.e. the received IFs, based on the states identified in the BCM (Figure 3.11).

As the switch's basic call handling is modeled with high-level finite state machines (FSMs), this model is also called *basic call state model* (BCSM) (ITU Recommendation Q.1214; ETSI TC-TR NA-60502). The BCSM identifies the logical points in basic call processing where the IN service logic located in the SCF is permitted to interact with basic call control capabilities provided by the switch. The BCSM identifies (only) those states and events that need to be made visible to IN service logic. Thus, the BCSM represents a projection of the BCP SIB of the GFP , described in section 3.1.1.2. The BCSM is built from three basic components (see Figure 3.11). These are:

1.  *Points in call* (PIC) provide an external view of a call-processing state or event to IN service logic. PIC are vendor independent, providing a standardized view of call-processing behavior. A PIC is characterized by means of entry event(s), exit event(s), actions performed within the PIC, and information available at the end of the PIC.

2.  *Detection points* (DPs), also referred to as *trigger check points*, are placed between the PICS, a DP being associated with a particular PIC.

DPs identify specific points in basic call processing at which specific events are detected and made visible to IN service logic, allowing for the transfer of control. The basic call processing *may be* suspended at a DP while waiting for instructions from IN service logic. Whether or not IN service logic will be invoked depends on the consultation of specific trigger conditions. If the trigger conditions at a DP are satisfied, then IN service logic processing will be initiated, otherwise (basic) call processing continues as if there had been no DP. Basically two types of DPs are distinguished in this context: *trigger detection points* (TDPs) are statically armed and used for initiating a service

Table 3.6 Mapping of GFP points of initiation and DFP detection points

| Point of initiation | Detection point |
| --- | --- |
| Call originated | Orig_Attempt_Authorized |
| Address collected | Collected_Info |
| Address analyzed | Analyzed_Info |
| Call arrival | Term_Attempt_Authorized |
| Busy | O_Called_Party_Busy |
| | T_Called_Party_Busy |
| | Route_Select_Failure |
| No answer | O_No_Answer |
| | T_No_Answer |
| Call acceptance | O_Answer |
| | T_Answer |
| Active state | O_Mid_Call |
| | T_Mid_Call |
| End of call | O_Abandon |
| | T_Abandon |
| | O_Disconnect |
| | T_Disconnect |

logic program, whereas *event detection points* (EDPs) are dynamically armed and are used to report the occurrence of a specific event to an already running service logic program.

3. *Transitions* indicate the normal flow of basic call processing from one PIC to another.

As the BCSM represents a projection of the BCP SIB, it is important to note that there are close relationships between the BCSM's DPSs and PIC and the BCP SIB's POI and POR, in other words a BCP POI corresponds to a particular BCSM DP, whereas a BCP POR can be mapped onto one or more BCSM PIC or DPs. The mapping of POI and POR onto corresponding DPs/PIC is given in Tables 3.6 and 3.7.

The current view of network operators and standards bodies is that there are two separate sets of basic call processing logic in the switching network (Figure 3.12), which are closely related:

1. one originating call model supporting the call's originating side (i.e. calling party) modeled by the *originating basic call state model* (O_BCSM); and

2. one terminating call model for the call's terminating side (i.e. called party) modeled by the *terminating basic call state model* (T_BCSM).

It has to be stressed that both sides of the call state model are active within a given node, with even an intraswitch call requiring both call state models.

Table 3.7 Mapping of points of return and detection points/points in call

| Point of return | Detection point/point in call |
|---|---|
| Continue with existing data | Several DPs — return to the same DP from which service logic was launched |
| Proceed with new data | Several PIC — return to the PIC specified by the service logic |
| Handle as transit | Analyze_Info or routing and alerting PICs |
| Clear call | O_Null or T_Null PICs |
| Provide call party handling | Several DPs — return to the same DP from which service logic was launched |
| Initiate call | Analyze_Info or routing and alerting PICs |

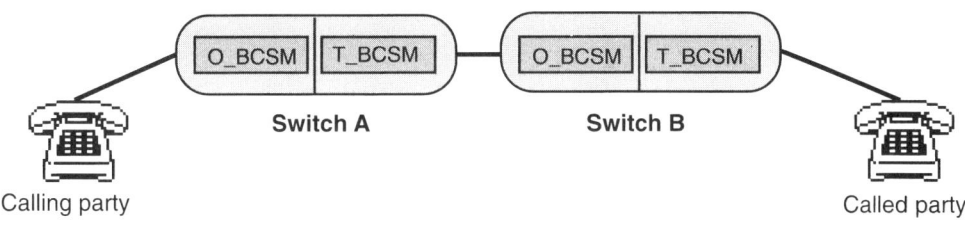

**Figure 3.12** Separation into originating BCSM and terminating BCSM.

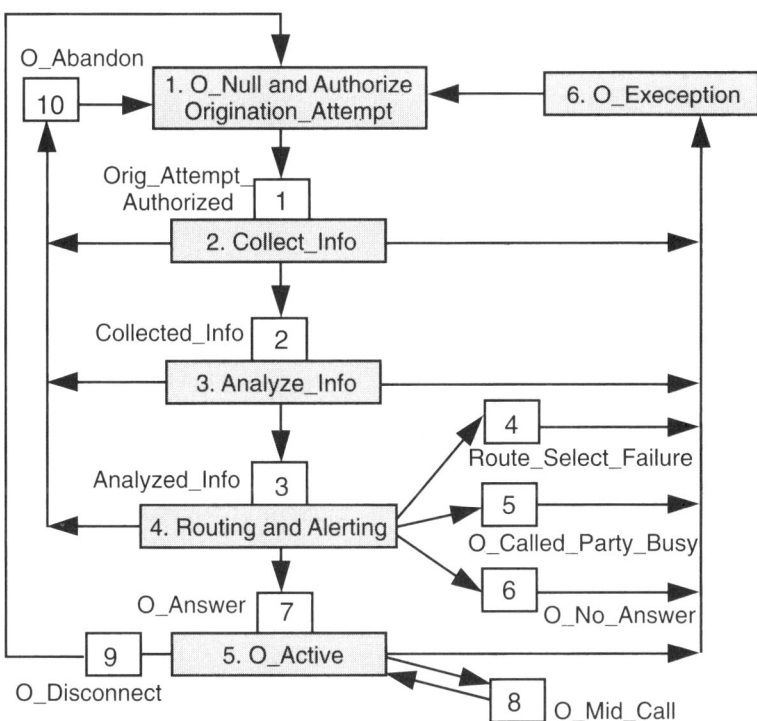

**Figure 13.13** Originating BCSM for CS-1.

Figure 3.13 depicts an example of an originating BCSM (O_BCSM) for CS-1, identifying for a given call six PIC and a corresponding set of 10 DPs.

Figure 3.14 indicates the corresponding terminating BCSM (T_BCSM), comprising four PIC and seven DPs. Note that these models are closely inter-related, that is O_BCSM's PIC 'Routing and Alerting' activates the T_BCSM (i.e. 'T_Null and Authorize Termination Attempt' PIC), and the 'T_Answer'

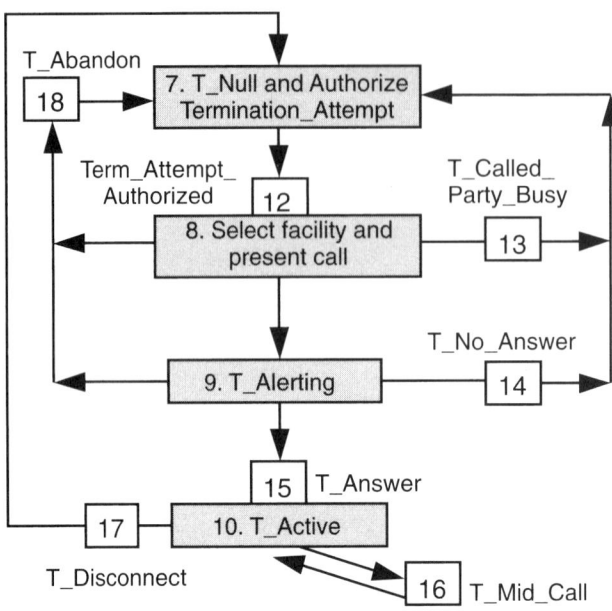

**Figure 3.14** Terminating BCSM for CS-1.

transition causes an 'O_Answer' transition in the O_BCSM. For details on these relationships readers are referred to ITU Recommendation Q.1214.

In summary it has to be recognized that the BCM and in particular DPs provide the basis for passing call control from the switch to an external IN service. In the next section will describe a typical call-processing scenario to illustrate the use of BCM and the DPs.

*A typical call-processing scenario*

We have learned that IN services are implemented within the DFP by distributed service logic, in other words for each IN service there must be a corresponding service logic program in at least one SCF, with the service data residing in the SDF. The service logic program interacts with service data, i.e. the SCF interacts with the SDF, for service execution. To make the IN service available within the network, the SSFs must include the corresponding service trigger information. This means that the exchanges have to be 'taught' about the remotely located IN service.

Furthermore, it must be emphasized that each IN service has a specific IN service number, which consists basically of two parts: a service access code, which identifies the service uniquely; and a subscriber-specific number, which identifies a particular service subscriber. The service access code is used to identify the service logic program, i.e. the SCF hosting this, whereas the subscriber-specific number identifies the subscriber data for that service (sometimes referred to as 'subscriber profile') in the SDF.

Putting the pieces together, IN call processing occurs as follows. The switch, i.e. the CCF/SSF, hosts a trigger table, which is partitioned according to the number of DPs identified in the BCSM. Each table entry contains, besides the trigger type, e.g. TDP or EDP, trigger criteria (sometimes also called 'DP criteria') indicating what conditions must be met (e.g. called number, calling number, line busy) and how to process the trigger if armed and active. (Note that each IN service will be identified by a specific service access number to allow IN service detection within the network, for example Freephone numbers start with '800' in the US and '130' in Germany.) Additionally, each trigger entry contains the appropriate SCF routing information. In particular, this routing information is important to identify a specific service logic program to which call processing control should be passed when the SSF detects an active trigger.

During call processing the CCF/SSF runs through the BCSM and examines at the DPs the associated part of the trigger table in sequence. This means that the SSF has to check if IN service trigger information, i.e. TDPs or EDPs, is armed. If the trigger is disarmed, the SSF will continue with basic call processing until it reaches the next DP. If the SSF detects that the trigger is active, then the BCSM execution is suspended and control is passed to the appropriate IN service logic program located at the SCF. Thus, the CCF/SSF generates, according to the identified DP (and PIC), a corresponding IF to the SCF. This is illustrated in Figure 3.15.

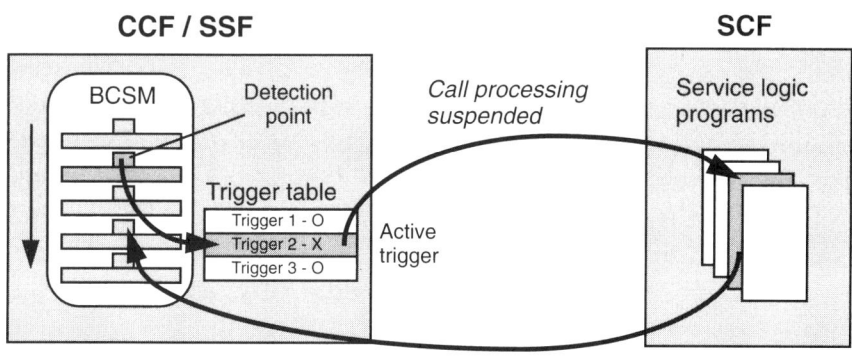

**Figure 3.15**   Detection of IN service calls in the switch.

On the receipt of this IF, the SCF determines the SSF conditions and directs the query to the corresponding service logic program. Note that, depending on the state of the call model and the type of DP, the appropriate service logic program has to be invoked or is already running. However, the service logic program at the SCF may generate new information for handling the call, e.g. through interaction with the SDF via corresponding IFs.

In order to pass the resulting instruction/information to the SSF, a corresponding response IF is generated by the SCF and sent to the SSF. Hence, after service logic program execution, the control is passed back to the CCF/SSF with the instruction that CCF/SSF should resume call processing at a particular PIC. (Note that the interactions between all these IN functional entities are achieved by corresponding IFs.) Based on this information, the CCF/SSF resumes the (basic) call processing until it arrives at the next DP in the call model, where it has to check again for an active trigger. Note that, depending on the functionality of a given IN service, multiple interactions between the CCF/SSF and the SCF may occur.

Let us consider a Freephone service call as an example. As for most IN services, the suspension of the basic call process happens at the 'collected info' DP (compare with Figure 3.13) which is located at the 'analyze info' PIC. Here the CCF/SSF checks the corresponding trigger table entry, which contains a number of service access codes. If the called number is a Freephone number, i.e. the number contains a Freephone access code, and this access code is contained in the trigger table entry, then the trigger will fire. The call control will be passed by the 'initial detection point' IF to the SCF, which is given by the second part of the trigger table entry.

After execution of the corresponding Freephone service logic program (which has performed a number translation based on the subscribed service features), the SCF will pass the destination number to the SSF. This will be done by means of the 'connect' IF. On reception of this IF, SSF continues basic call processing at the 'routing and alerting' PIC. Hence, the call will be established to the called Freephone number. Note that in case of the Freephone service it is important to monitor the call charges in order to place these later on the account of the Freephone service subscriber. Thus, the SSF has to interact again with the SCF after call completion. This means that at the 'O_Disconnect' DP another interaction between SSF and SCF will occur to transfer the call charges.

### 3.1.2.2 CS-1 physical plane

The physical plane is the lowest plane of the INCM and defines the 'real' physical IN architecture. In the next two sections we look at the physical IN elements and the basic protocol used for the communication between these physical elements, known as IN application protocol.

*Physical elements*

Each of the IN FEs described above can be mapped to a physical entity of an IN-structured network. The physical IN architecture is defined in the IN physical plane (ITU Recommendation Q.1215; ETSI ETS 300348), indicating possible mappings of the DFP functional entities onto *physical entities* (PEs). A complete FE must be implemented within one single PE. Different FEs can be implemented in the same or different PEs according to different characteristics of the underlying network technology and service-specific access requirements.

These PEs defined are similar to those elements we have already introduced in section 2.4. However, additional entities have been identified. A typical

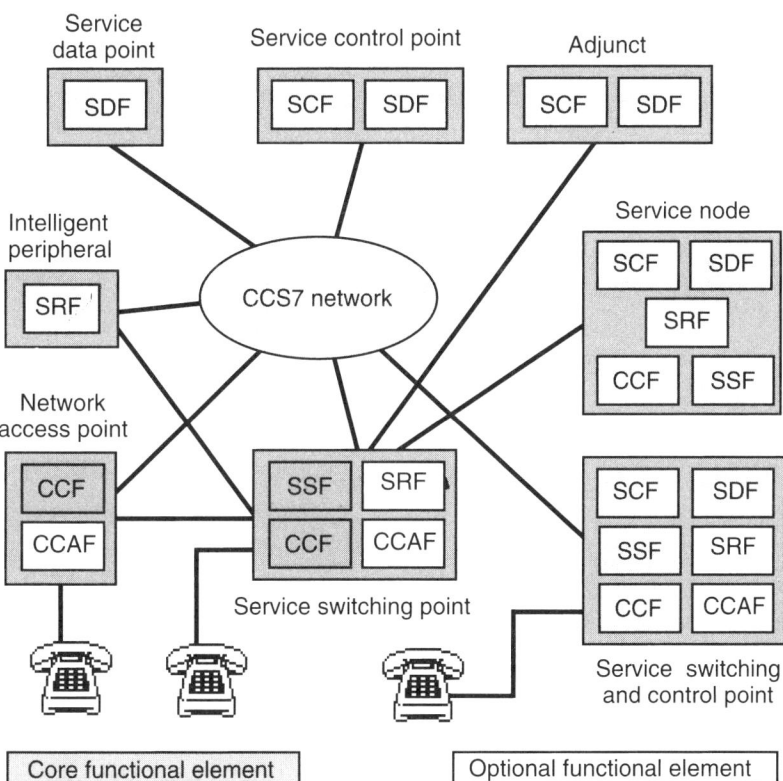

**Figure 3.16** IN physical elements (call-related elements only).

physical IN-structured network will consist of some or all the following components (Figure 3.16).

- A *network access point* (NAP) contains only an CCAF and an CCF and thus cannot be really considered as an IN physical element. It is only introduced to enable early and ubiquitous deployment of IN services into existing networks. A NAP cannot directly communicate with an SCP. However, it is able to identify that a call requires IN processing and will route the call to an SSP for IN service processing.

- A *service switching point* (SSP) provides the service switching function (SSF) and the connection control function (CCF). It provides central switching resources, as well as signaling and bearer interfaces to other network components. The SSP contains detection capabilities to detect requests for IN services. When the SSP has subscribers directly connected to it, it also provides the call control agent function (CCAF). In addition, an SSP may contain a specialized resource function (SRF).

- A *service control point* (SCP) provides and contains a service control function (SCF) and a service data function (SDF). Hence, the SCP contains the service logic programs and the data that are used to provide IN services. Note that multiple SCPs may contain the same service logic programs and data to improve service performance and reliability. The SCP is connected to SSPs through the signaling network. In addition, it may access an external service data point (SDP) either directly or via the signaling network. Optionally, the SCP can also be connected to an intelligent peripheral (IP).

- A *service data point* (SDP) provides the platform for a stand-alone service data function (SDF), and thus contains customer and network data which is accessed during the execution of a service.

- An *intelligent peripheral* (IP) is the physical entity responsible for the implementation of the specialized resource function (SRF). Hence, it enables flexible information interactions between a user and the network. The IP may directly connect to one or more SSPs, and/or may connect to the signaling network. Control of the IP can be either directly by an SCP/adjunct (via a signaling link) or via an SSP relay function. An SCP or adjunct can request an SSP to connect a user to a resource located in an IP that is connected to the SSP from which the service request is detected.

- A *service switching and control point* (SSCP) is a combined SSP and SCP in a single node. Functionally, this entity contains an CCAF, CCF, SSF,

SCF, and SDF. The interconnection between the CCAF/CCF/SSF and the SCF/SDF is proprietary and closely coupled. However, it provides the same service capabilities as an SSP and SCP separately. The interfaces between the SSCP and other physical elements are the same as the interfaces between the SSP and other physical entities. The motivation for this entity is to achieve IN service in the case of limited CCS7 penetration.

- An *adjunct* is functionally equivalent to an SCP (i.e. it contains the same functional entities), but is directly connected to one or more network switches (i.e. SSPs). The basic difference between an SCP and an adjunct is that communication between an SSP and the adjunct is supported by a high-speed interface in contrast to a CCS7 link! This arrangement may result in differing performance characteristics for an adjunct and an SCP. Like the SSCP, an adjunct is attractive in case of limited CCS7 penetration.
- A *service node* (SN) combines an CCF/SSF, SCF, SDF, and SRF. It can control IN services and engage flexible information interactions with users. An SN communicates directly with one or more external SSPs, each with a point-to-point signaling and transport connection. In a manner similar to an adjunct, the SCF in a SN receives messages from the SSP, executes service logic programs, and sends messages back to the SSP. The SRF in a SN enables it to interact with users in a manner similar to an IP. Also, a SN is attractive in case of limited CCS7 penetration.

Table 3.8 provides an overview of the possible mapping of functional IN elements onto physical elements. Note that the network access point (NAP) is not included, as it represents no real IN physical element.

In addition to the call-related physical IN elements described above, three other physical elements can be identified. The *service management point* (SMP) provides the *service management function* (SMF), whereas the *service management agent point* (SMAP) implements the *service management agent function* (SMAF). The SMP interfaces with virtually every other physical component of the intelligent network. As the SMP and the SMAP are not fully developed within CS-1, the implementation of the functions could vary. An additional *service creation environment point* (SCEP) supports the *service creation environment function* (SCEF). Like the SMP and SMAP, the SCEP is not fully defined within CS-1, thus the implementation of the functions could vary.

Looking at the implementation of a real IN, the number of each entity deployed in the network depends on the geographical coverage of the network,

Table 3.8 Mapping of functional entities onto physical entities. Dark symbols indicate mandatory FEs in PEs, light symbols indicate optional FEs in PEs

| Physical elements | Functional elements | | | | | | |
|---|---|---|---|---|---|---|---|
| | Service switching function | Service control function | Specialized resource function | Service data function | Service management function | Service management agent function | Service creation environment function |
| Service switching point | ● | ○ | ○ | ○ | | | |
| Service control point | | ● | | ○ | | | |
| Intelligent peripheral | ○ | | ● | | | | |
| Service data point | | | | ● | | | |
| Adjunct | | ● | | ● | | | |
| Service node | ● | ● | ● | ● | | | |
| Service switching control point | ● | ● | ○ | ● | | | |
| Service management point | | | | | ● | ○ | ○ |
| Service management agent point | | | | | | ● | |
| Service creation environment point | | | | | | | ● |

the anticipated call volume for the IN services to be provided, CCS7 deployment, and the level of deregulation.

Implementing SSFs and SCFs/SDFs in separate physical entities, that is implementing dedicated SSPs and SCPs, is probably the most elegant way to establish an IN. In this case one will usually find numerous SSPs and quite a few SCPs controlling the SSPs. Additionally, there will be probably only one or two service management points, often referred to as a service management system, for a complete IN.

However, this scenario necessitates the availability of CCS7 network and in particular INAP on top of it. Unfortunately, this is not commonly the case today. Thus, the best way to implement an IN without CCS7 availability or limited CCS7 penetration, as has been the case for most countries in recent years, is to allow the SSF and the SCF/SDF to be physically present in the same network node, or to connect these entities by a direct link. In order to support these scenarios the service switching and control point, adjunct and service node have been defined. This method of implementation does not require CCS7, as the communication between the SSF and the SCF is achieved via an internal or proprietary interface. In this way an IN can be introduced at a very early stage.

Coming back to an IN implementation based on dedicated SSPs, SCPs, and IPs, it is necessary to take a look at how the communication between these entities is achieved via the CCS7 network. Thus, in the next section we focus on the *IN application protocol* (INAP).

*IN application protocol*

As outlined above, a complete DFP functional entity must be implemented within one physical entity; different functional entities, however, can be implemented in the same or different physical entities. If two of these functional entities are implemented in remote physical entities, they need to communicate to achieve an IN service. These information flows defined in the DFP are implemented in the physical plane through a standardized OSI application layer protocol, known as the *IN application protocol* (INAP) (ITU Recommendation Q.1218), also referred to as *'core INAP'* within ETSI ETS 300374-1.

To enable the implementation of IN CS-1 service and service features, INAP supports the interactions between the following four functional entities: service switching function, service control function, specialized resource function, and service data function. However, in the current INAP version the interactions with the SDF are not yet defined. Note that, alternatively, the revised CS-1 recommendations foresee the use of the directory access protocol (DAP) (X.500) for interactions between SCF and SDF. Consult ITU Recommendation Q.1218 for details.

In general, INAP makes use of some general *application service elements* (ASEs), such as *transaction capabilities* (TCs) and *remote operations service elements* (ROSEs). It defines 25 INAP-specific ASEs. These additional ASEs are based on the criteria for grouping operations into ASEs, which include functional distribution, modular reuse, and future evolution.

As INAP aims for a modular structure, ASEs can be grouped in so-called 'application contexts', where an application context consists of a combination of ASEs and the relationship between these ASEs. An application context is typically a subset of the total INAP and specifies that portion of the protocol needed in the communication between two types of IN functional entities. For example, between the SSF and SCF functional entities a possible application context would bind for a given transaction the ASEs for basic call processing, charging, status reporting, etc. Similarly, other application contexts could be established between the SCF and the SRF for user interaction, between the SCF and SDF for data retrieval, and so forth. This is depicted in Figures 3.17 and Figure 3.18.

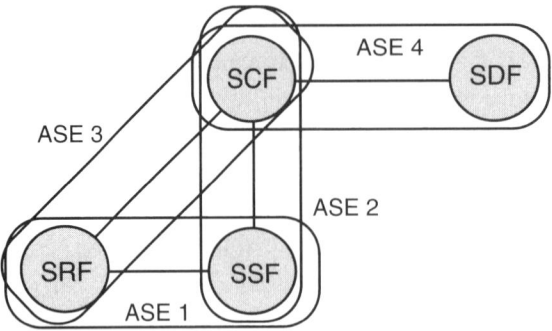

**Figure 3.17** Different application service elements between different IN functional entities.

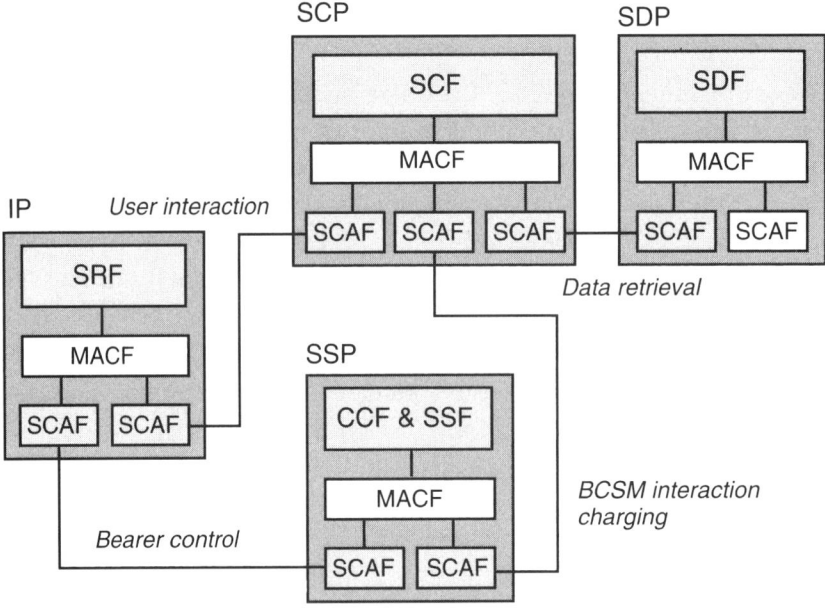

**Figure 3.18** Different INAP application service elements between IN physical elements.

The INAP protocol architecture can be described as follows. A physical entity has either a single interaction or multiple coordinated interactions with other physical entities. In the case of a single interaction, a *single association control function* (SACF) provides a coordination function in using a set of ASEs. In the case of multiple coordinated interactions, a *multiple association control function*

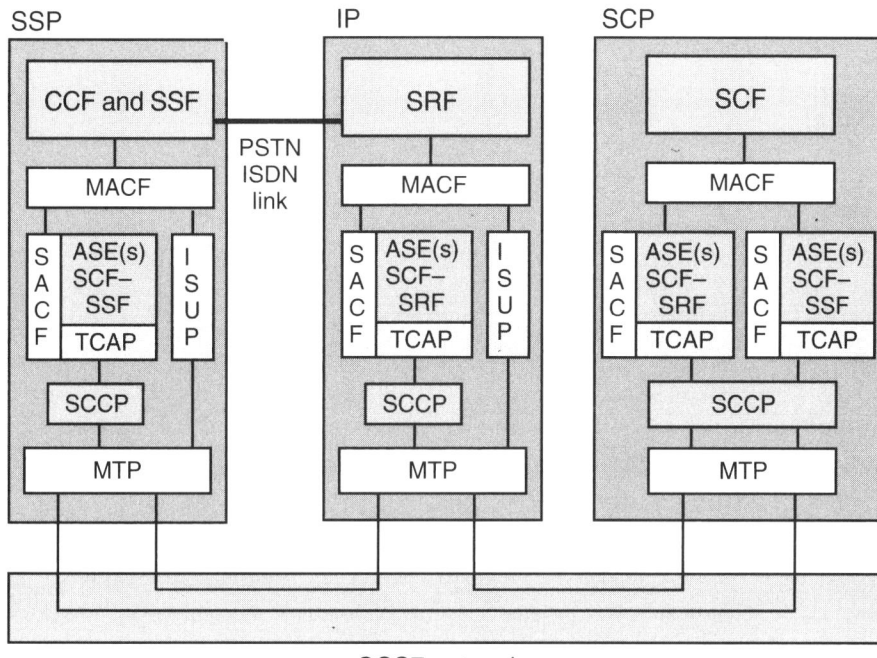

**Figure 3.19** INAP scenario for separate SSP, IP and SCP based on CCS7.

(MACF) provides a coordination function among several sets of ASEs, where each ASE supports one or more operations. One of ASEs is the transaction capabilities application part (TCAP) in the case of CCS7. Figure 3.18 illustrates the use of different ASEs between different physical entities.

At the moment INAP makes use of existing signaling protocol standards and platforms, such as CCS7's TCAP, which is based on ISO's 'application layer structure' and includes ITU's ROSE (Figure 3.19). Thus, the IFs between the IN client–server pairs, e.g. SSF–SCF, are achieved in INAP by using remote operations. INAP supports any mapping of IN functional to physical entities. Generally, the operations defined within INAP are specified using the abstract syntax notation 1 (ASN.1) formalism. These correspond to the operations and IFs between the functional entities (ITU Recommendation Q.1214). Examples are:

- analyze information,
- apply charging,
- assist request instructions,

- collect information,
- connect,
- connect to resource,
- continue,
- release call,
- play announcement,
- prompt and collect user information.

## 3.1.3 Putting the pieces together – service creation in CS-1

To summarize the basic points of the previous sections and the related CS-1 standards, this section illustrates how 'new' or tailored IN services can be constructed in an IN environment.

In general, IN service modeling and implementation within CS-1 is based on the INCM following a top-down approach. This means that the definition of a new IN service starts with a description of the envisioned service without reference to distribution of functionality in the network. Thus, the service has to be decomposed into according services features in the service plane. All identified service features have to be mapped onto corresponding SIBs within the GFP.

Given that there already exists a catalog of services, service features, and SIBs, it is necessary to check whether the planned service functionality can be achieved by already existing service features and in particular existing SIBs. In particular, the SIBs and the BCM (i.e. the identified POI with the BCP SIB) are of prime importance in this context, as these provide the basis for service feature implementation.

In the case that the existing set of SIBs is not sufficient, an enhancement of already existing SIBs or even a new SIB definition and specification may be required. Moving from the GFP to the DFP, an enhancement of the BCM in terms of new DPs and PIC definitions may be required in case the BCP SIB has been modified. If new SIBs have been defined in the GFP, it is necessary to check whether the existing set of FEs, FEAs, and IFs is sufficient to support the distributed implementation of the envisioned SIB functionality in the DFP. If this is not the case, new FE and FEAs and/or additional IFs may have to be specified. Note that this may result also in corresponding enhancements of the IN application protocol within the physical plane.

However, it must be stressed that the definition of call model enhancements and/or new SIBs and the resulting enhancements in the DFP and the physical plane should be considered as a subject for IN standardization, as only the uniformity in call model and IFs use and the corresponding uniformity in the interfaces of physical IN elements allows for multivendor IN equipment. IN capability set 2 standardization, as described in the section 3.2, is the best example of how new benchmark services result in corresponding enhancements of the INCM plane. This means that, staying within the scope of CS-1 standards, only the existing SIBs, including the existing BCM, should be used to provide new services (compare with the beginning of this chapter).

However, coming back to our initial consideration of the service creation process, the service designer (who usually works for the service provider, as service creation is closely related to the used IN platform) is basically modeling the arrangement of SIB chains (sometimes referred to as 'service scripts'), including their interaction with the BCP SIB. In addition, the service designer has to model the required SIB input data, comprising SSD and CID, which supports each execution instance of an IN service.

Once the global service logic, i.e. the SIB chain(s) and its relations to the BCP, has been defined, it has to be mapped onto the distributed IN architecture. The mapping of global service logic onto distributed service logic, that is mapping of an SIB chain onto the distributed service components in the DFP, i.e. the service logic program in the service control function, the service data in the service data function, the service trigger information in the service switching function, and optional data/functions in the specialized resource function, is a complex task that is not currently standardized at this time and depends on the particular IN platform used.

The development of 'new' or tailored IN services, including the establishment of global service logic, i.e. SIB chains, and the generation of the corresponding distributed service logic, comprising service logic program, service data template, and service trigger information, will usually be supported by so-called 'service creation environments'. Basically, these service creation environments offer graphical–user interfaces supporting the dynamic combination of SIBs and produce as output the required distributed service logic. As currently most vendors' IN platforms make use of proprietary sets of SIBs (i.e. the SIBs defined in different platforms differ in their functionality and number), each IN platform offers its proprietary service creation environment. Furthermore, the service creation environment is not yet specified within CS-1 standards. However, this situation may change in the long term with the progress of IN standardization.

Closely related to the service creation process is testing and validation of the generated distributed service logic. Thus, a service creation environment is mostly closely coupled with a service testing and validation environment, which allows simulation of the new service components within a 'real' IN platform. Here the 'basic service software' can be customized to meet customer-specific parameters (e.g. call distribution parameters, such as number of destinations to be reachable, time- and origin-dependent routing value). This includes software testing in general, as well as simulating the distribution of distributed service logic within the network (see below).

After service testing and validation, the new/tailored service, i.e. its components, has to be introduced/installed into the real physical IN network before it can be provided to the service subscriber and used by end users. This is a service management activity and will be performed by the service management system, which makes use of the output data of the service creation environment. The distribution and/or replication of service logic and service data onto existing SCPs and SDPs, as well as the introduction of the service trigger information into the switches, i.e. SSPs, depends on the desired geographical coverage and the expected call volume of the service. Figure 3.20 illustrates the whole service creation and installation process.

This means that if the new service should be only offered within a limited geographical area with low expected call volume, it is sufficient to make use

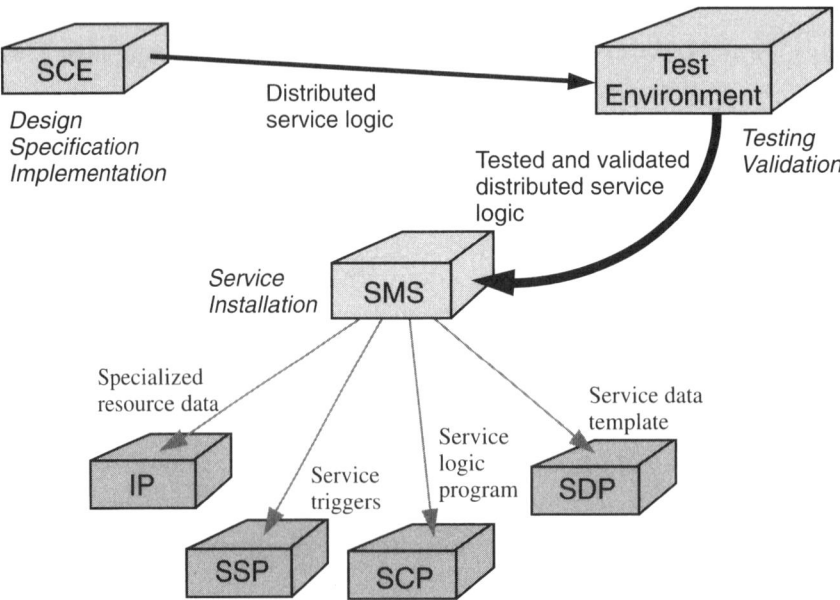

**Figure 3.20** Service creation and installation.

of a single SCP and SDP, installing the service trigger information only on those SSPs serving the desired area. Otherwise, replicated service logic and data will be installed at multiple SCPs and SDPs and in all SSPs to allow for highly parallel service use.

It must be stressed that service creation, testing, and validation is a very important research issue in the field of telecommunications. Many international research projects within ESPRIT, EURESCOM, and RACE are involved in the development of appropriate tools for service creation for open service environments. Note that the issue of solving service feature interaction problems is also an essential part of the service creation process. A good example of a service creation environment is given in Carvalho *et al.* (1994).

## 3.1.4 Examples of IN services and their implementation

In the next subsections we provide some examples of IN services defined within capability set 1, which are already provided or will be provided soon by IN platforms in most countries. These are:

1.  Freephone
2.  premium rate
3.  televoting, and
4.  card calling services.

Note that the following service descriptions should be regarded as examples only. Real implementations of each service could vary according the service definition, service feature implementation, IN platform capabilities, and CCS7 penetration.

### 3.1.4.1 Freephone

Freephone, also known as 'Green Number' or 'Toll Free' service is probably the best-known IN service, and it is one of the commercially most interesting IN services. Basically, Freephone can be considered as a basic driver for IN development in North America in the 1980s. The Freephone service has been available in the traditional telecommunication networks for many years (Chapter 2), but today it is one of the first IN services to be introduced in nearly every country.

Each IN service must be identified by a specific service access number to allow IN service detection within the network; Freephone numbers start with '800' in the US and with '130' in Germany. In addition, this service number concept allows potential service users to recognize that they are using a telecommunications service with specific characteristics, such as the fact that a call will cost nothing (e.g. a Freephone service) or cost more than usual (e.g. a premium rate service).

*Basic service functionality*

With the Freephone service the call charges are allocated to the called party, representing the service subscriber. This means that calling users (i.e. service users) do not have to pay the call charges when calling a Freephone number. In addition, the service subscriber is reachable with only one single number, i.e. the Freephone number, regardless of the number of existing terminating lines. A caller who dials a Freephone number is connected to the terminating line specified by the service subscriber. The Freephone service could be regarded as an enhanced 'universal access number' service in that it allows subscribers (usually but not necessarily business companies using the service for promotion purposes) with offices in several locations to be reached on one number common to all offices. To which office a call should be routed depends on several parameters, particularly the area where the call originated. Examples of businesses subscribing to the Freephone service include fast food delivery restaurants, hotels, airlines, and car rental stations.

The basic service functionality is defined by the two core service features 'one number' and 'reverse charging' (section A.3).

1.  The *one number* service feature allows a subscriber with two or more terminating lines in any number of locations to have a single telephone number. The subscriber may specify which calls are to be terminated on which terminating lines based on the area where the calls originate.
2.  The *reverse charging* service feature allows the service subscriber to be charged for the entire cost of a call.

*Optional service functionality*

The basic Freephone functionality outlined above can be enhanced by various optional service features to meet the demands of specific business customers. In accordance with the table in section A.3, the following optional service features may be incorporated within a specific Freephone service realization:

*   authentication
*   call distribution
*   call forwarding conditional
*   call gapping

- call limiting
- call logging
- call queuing
- customer profile management
- customer recorded announcement
- customized ringing
- destination user prompter
- mass calling
- originating call screening
- originating user prompter
- origin-dependent routing
- time-dependent routing.

For a description of all of these service features readers are referred to section A.2. Considering these service features, it becomes clear that Freephone is a service designed for business companies expecting high call volumes and having different offices/branches. Thus, service features such as call queuing, call limiting, and mass calling, as well as call distribution, origin-dependent routing, and time-dependent routing can be incorporated within a particular Freephone service implementation.

*Service example*

In the following example we focus on the incorporation of two optional service features, namely time-dependent routing and originating user prompter, in order to illustrate how enhanced or tailored versions of a Freephone service can be provided. The description of the above two service features is as follows.

- *Time-dependent routing* allows the served user to apply different call treatments based on time of day, day of week, day of year, holiday, etc.
- *Originating user prompter* enables the prompting of the calling party with a specific announcement. Such an announcement may ask the calling party to enter an extra number (e.g. through dial tone multiple frequency) or a voice instruction that can be used by the service logic for continuing to process the call.

With the 'time-dependent routing' service feature, the service subscriber can define to which terminating line, e.g. physical destination, calls to its Freephone number will be routed. This may be useful for companies offering an emergency office during the night hours; in this case calls between 18.00 and 08.00 hours would be routed to the emergency office, whereas calls between 08.01 and 17.59 hours would be routed to the usual offices. In addition, calls at weekends and public holidays could be handled in a similar way.

The originating user prompter service feature is attractive for those service subscribers who want to identify, by an automated dialog with the user, 'the callers' purpose so that they can be connected to the appropriate terminating line. For example, a car company may have a single Freephone number for all its departments, comprising new car sales, used car sales, car repair, spare parts, etc. On dialing this number callers would be asked to enter a '1' if they want to buy a new car, to enter a '2' if they are interested in an used car, to enter a '3' if they want to book their car in for repair, etc.

*Distributed service processing*

Having introduced the functionality of a Freephone service, we will now outline the implementation within the IN architecture. We assume that the service features involved in a Freephone service have been decomposed into appropriate SIB chains (see sections 3.1.1.2 and A.4) and that the corresponding distributed service logic has been created (see section 3.1.3). This means that there is service trigger information for the service switching function, a service logic program for the service control function, service data for the service data function, and optional specialized resource data (e.g. customized announcements) for the specialized resource function.

In the following we assume that SSPs contain the service switching function, an SCP contains both the service control function and the service data function, and an optional intelligent peripheral (IP) implements the specialized resource function. In addition, switches implement basically the connection control function, where specific switches also host an SSP.

*Implementation of the basic Freephone service*

Figure 3.21 illustrates a basic Freephone implementation. The call processing is described below.

1.  The calling party dials the Freephone number (e.g. '130-12345').
2.  The switch recognizes during basic call processing that the call is a Freephone call. It suspends basic call processing and the SSP sends, based on the information in the trigger table, an INAP query to the corresponding SCP, which contains among other data the calling and the called line identification. On receipt of the query, the SCP starts

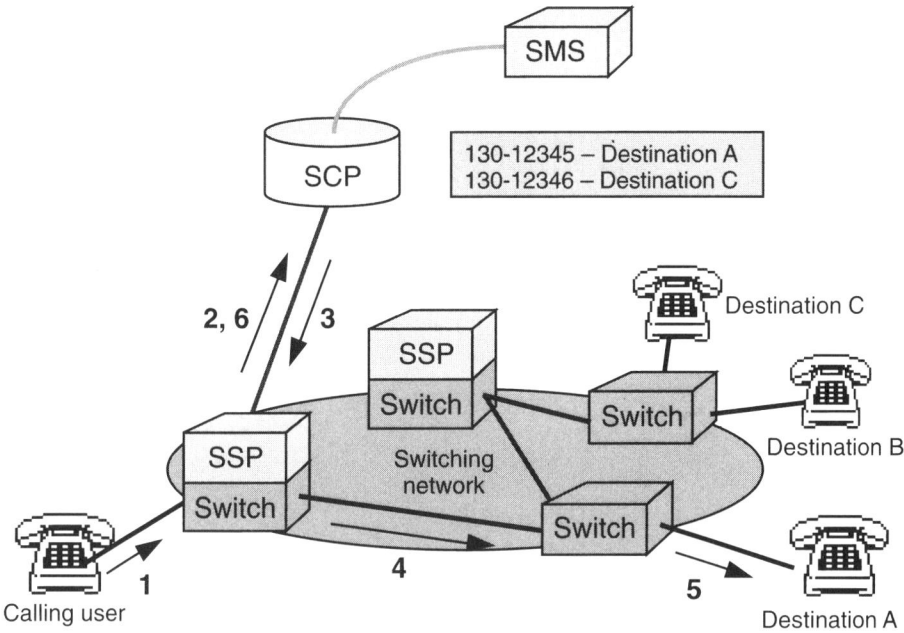

**Figure 3.21** Basic Freephone service processing.

the corresponding service logic program and translates by means of corresponding subscriber service data the called number (representing the Freephone number '130-12345') into the corresponding destination number (e.g. 'destination A').

3. The SCP sends the destination number back to the SSP. In addition, the SCP instructs the SSP to monitor the call charges and pass the call charge record after the call to the SCP to place these charges on the subscriber's bill.

4. On receipt of the destination number, the SSP instructs the CCF to resume call processing with the new destination address. The switch routes the call over the public network to the destination switch and transmits the connection sequence.

5. The destination switch establishes a connection to the called party, i.e. 'destination A'. The call between the calling and the called party is established.

6. After the call is terminated, the SSP sends a message to the SCP in order to inform the SCP about the call charges. The SCP adds the charges to the subscriber account.

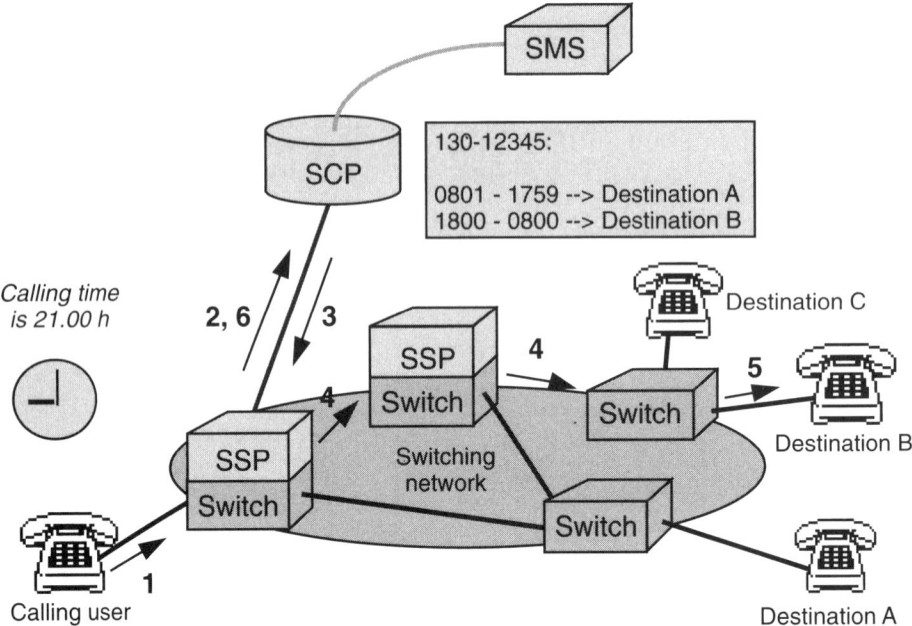

**Figure 3.22**  Freephone service processing with time-dependent routing.

*Implementation of Freephone service with time-dependent routing*

Figure 3.22 illustrates an implementation of a Freephone service incorporating time-dependent routing. The call processing is as follows.

1. The calling party dials the Freephone number (e.g. '130-12345').
2. The switch recognizes during basic call processing that the call is a Freephone call. The SSP sends an INAP query to the corresponding SCP. On receipt of the query the SCP starts the corresponding service logic program. In addition to the previous scenario, this time the service logic program and the service data are enhanced by the ability to route the call in accordance with the time, day of week, etc. Thus, the service logic program compares the call time with the predefined routing table to determine the appropriate destination number.
3. The SCP sends the destination number back to the SSP ('destination B'). In addition, the SCP instructs the SSP to monitor the call charges and pass the call charge record after the call to the SCP to place these charges on the subscriber's bill.
4. On receipt of the destination number the SSP instructs the switch to route the call over the public network to the destination switch.

5.  The destination switch establishes a connection to the called party, i.e. 'destination B'.
6.  After call termination, the SSP reports the call charges to the SCP. The SCP adds the charges to the subscriber's account.

*Implementation of Freephone service with originating user prompter*

Figure 3.23 illustrates an implementation of a Freephone service incorporating originating user prompter. The call processing is as follows.

1.  The calling party dials the Freephone number (e.g. '130-12345').
2.  The switch recognizes during basic call processing that the call is a Freephone call and the SSP sends an INAP query containing call information to the corresponding SCP. On receipt of the query the SCP starts the corresponding service logic program. This time the service logic program and the service data are enhanced with the capability to interact with the user in order to determine the destination number. Thus, the SCP identifies an appropriate user interaction device, i.e. an IP, which contains the desired announcements.

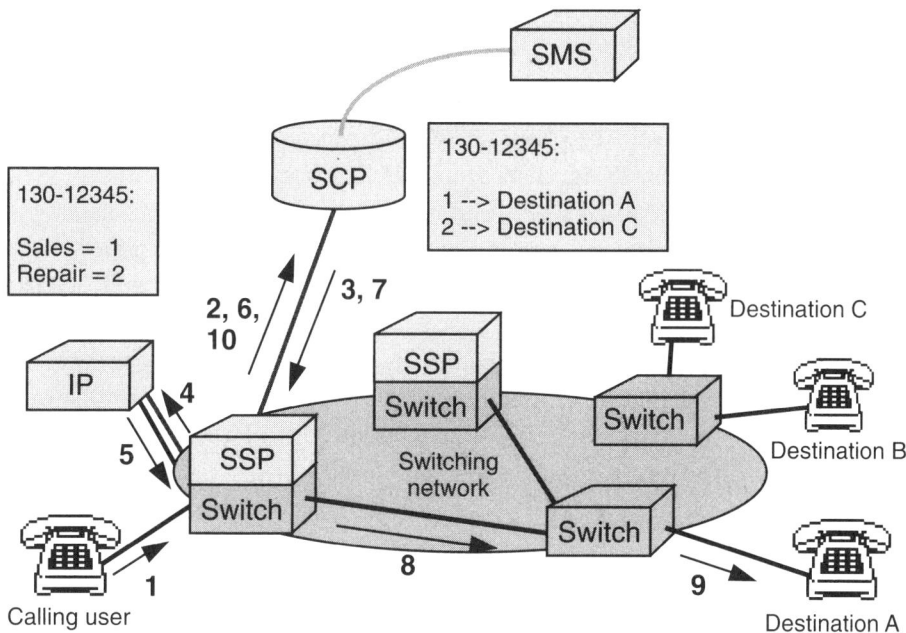

**Figure 3.23**   Freephone service processing with originating user prompter.

3. The SCP returns a routing number of an appropriate user interaction device, i.e. an IP, to the SSP.

4. The SSP instructs the switch to establish a connection with the IP. Furthermore, it asks the IP to start a dialog with the user. The IP interacts with the user by playing the desired announcements and collecting the response digits.

5. The IP returns the user selection to the SSP.

6. The SSP relays the user selection to the SCP.

7. The SCP examines the user selection. It instructs the SSP to release the connection with the IP and provides the SSP with routing and charging information for the call to be established.

8. On receipt of this information the SSP instructs the switch to release the IP connection and to route the call over the switched network to the destination switch. In addition, call charge monitoring is started.

9. The destination switch establishes a connection to the called party, i.e. 'destination A'. The call between the calling and the called party is established.

10. After call termination, the SSP reports the call charges to the SCP. The SCP adds the charges to the subscriber's account.

### 3.1.4.2 Premium rate

The premium rate service, also known as Kiosk, is an IN service with great commercial potential. This service provides the service subscriber, representing an external information service provider, with a premium rate number. Users who call this number are charged at a special rate for both the call and the information/service obtained by the call. The network operator collects the revenue and shares it with the service provider. Examples of premium rate services are weather or stock information, health consultation, or lonely hearts services.

In principle, this service is quite similar to the Freephone service, as the premium rate number is also a specific service number (e.g. '900' in North America and '190' in Germany) and special treatment of the call charges is provided.

*Basic service functionality*

Compared with Freephone calls, premium rate service calls are expensive, as the calling party has to pay, in addition to the basic call charges, an extra charge for the information service offered. These combined charges are collected with the telephone bill from the calling user, with the network operator passing the extra revenue to the premium rate service subscriber. The premium rate service subscriber is able to earn revenue for each call made to a premium rate number.

As with the Freephone service, the premium rate service subscriber, usually a company, is reachable with only one number, namely the premium rate number, regardless of the number of existing terminating lines.

The basic service functionality is defined by the two core service features 'one number' and 'premium charging' (section A.3).

- The *one number* service feature allows a subscriber with two or more terminating lines in any number of locations to have a single telephone number. The subscriber may specify which calls are to be terminated on which terminating lines depending on the area the calls originate.
- The *premium charging* service feature allows some of the revenue from a call to be returned to the called party, i.e. the service subscriber, acting in the role of a value-added service provider.

*Optional service functionality*

The basic premium rate functionality outlined above can be enhanced by various optional service features to meet the demands of specific business customers. In accordance with the table given in section A.3, the following optional service features may be incorporated within a specific Freephone service:

- call distribution
- call forwarding conditional
- call gapping
- call limiting
- call logging
- call queuing
- customer profile management
- customer recorded announcement
- customized ringing
- originating call screening
- originating user prompter
- origin-dependent routing
- time-dependent routing.

For a description of all of these service features readers are referred to section A.2.

*Service example*

In the following we consider an information service provider which provides the newest stock exchange information from all over the world. This information could be provided either via a usual dialog from person to person or via a tape playing the advice to calling users. The 'Stock Information' service is advertised under the premium rate number '190-12345'. Thus, users have to dial this service number to obtain the desired information. After call completion the call charges, comprising both basic call charges as well as additional surcharges for the premium rate service use, will be added to the calling user's account. The revenues collected from that service will be allocated to the service subscriber, i.e. the stock information service provider, periodically (e.g. monthly).

*Distributed service processing*

Figure 3.24 illustrates a basic premium rate service implementation. The call processing is described below.

1. The calling party dials the premium rate number (e.g. '190-12345').
2. The switch recognizes during basic call processing that the call is a premium rate call. Basic call processing is suspended and the SSP sends an INAP query to the corresponding SCP. On receipt of the query the SCP starts the corresponding service logic program and translates the called number (representing the premium number '190-12345') into the corresponding destination number (e.g. 'destination A') using the subscriber service data.
3. The SCP responds to the SSP with the destination number and billing information. The latter is used to apply a special charging treatment of that call and allows the extra charges to be paid by the calling party to be determined.
4. On receipt of the destination number the SSP instructs the switch to route the call to the destination switch.
5. The destination switch establishes a connection to the called party, i.e. the information service provider. Now the calling party may obtain the desired service from the information service provider.
6. After call termination, the SSP reports the call charges to the SCP. The SCP places both the basic call charges and the extra charges on the calling party's bill.
7. In addition, it forwards the extra charges, i.e. the revenues, to the premium rate service subscriber.

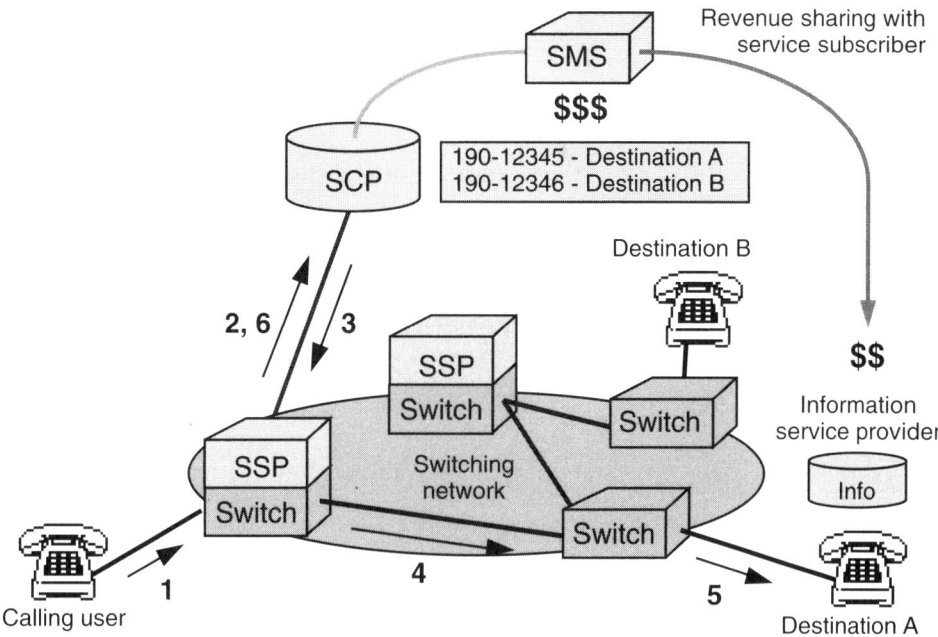

**Figure 3.24**   Basic premium rate service processing.

### 3.1.4.3 Televoting

Televoting is also a very popular service in many countries, and is one that has been available in the traditional telecommunication networks for many years. This service is becoming popular because of the increasing frequency with which opinion polls are conducted during radio and TV shows. The service allows the telephone network to be used for voting. Televoting numbers start with a specific service access number, e.g. '137' in Germany.

*Basic service functionality*

The televoting service allows the service subscriber to conduct a phone vote in which service users may chose between two or more options by telephone voting There are two methods of implementing the service: either callers are asked to ring a specific number depending on which option they want to vote for or they ring a unique number and, after prompting, give their choice by keyboard or by voice dialog. Calls to the televoting number are registered cumulatively and a recorded announcement is sent to calling users to confirm the registration of their vote. As with the premium rate service, the call charges may be somewhat higher than for a normal phone call. The service subscriber, i.e. the initiator of the voting, can query this cumulative register facility at any time by following the necessary authorization procedures.

In contrast to Freephone or premium rate services, subscription to a televoting service is usually limited to a short period, i.e. a few minutes or hours, during which the phone poll should be conducted.

The basic service functionality is defined by the 'mass calling' service feature (section A.3).

- The *mass calling* service feature allows processing of huge numbers of incoming calls generated by broadcast advertising or games.

*Optional service functionality*

The televoting subscription may include a number of additional service capabilities that allow the service to be customized according to specific subscriber needs. For example, service subscribers may want the televoting calls accepted or rejected based on call origin, time, calling line identity, etc. Furthermore, service subscribers may want some or all incoming calls to be routed to their installation(s) or rejected according to predefined criteria. Optionally, the call handling may be altered by the service subscriber.

Based on the table given in section A.3, the following optional service features may be incorporated within a specific televoting implementation (for a description of all of these service features readers are referred to section A.2):

- call distribution
- call gapping
- call limiting
- call logging
- call queuing
- customer profile management
- customer recorded announcement
- originating call screening
- originating user prompter
- origin-dependent routing
- time-dependent routing.

*Service example*

In the following we will look at a particular televoting service implementation, including the 'originating user prompter' service feature. This service feature is used to inform calling users that their vote has been counted. The service example is as follows. Within a TV show there is a talent competition. After some of the competitors have performed their acts, the viewers are asked to vote for one out of, say, eight competitors. In the first scenario the watchers should dial one out of eight different numbers, these numbers differing only in the final digit. For example, viewers voting for act number 1 would dial '137-12301', whereas those voting for act number would dial '137-12302', etc. Upon dialing such a number, calling users will be played a message indicating that their vote has been taken into account.

After the opinion poll, which has been limited to 30 minutes, the TV company queries the results of the phone vote from the service provider and presents the results to the audience.

*Distributed service processing*

In the first scenario the call processing looks as follows (Figure 3.25).

1. The calling party dials the televoting number (e.g. '137-12301').
2. The switch recognizes during basic call processing that the call is a televoting call and the SSP sends an INAP query containing call information to the corresponding SCP. On receipt of the query, the SCP starts the corresponding service logic program. Depending on

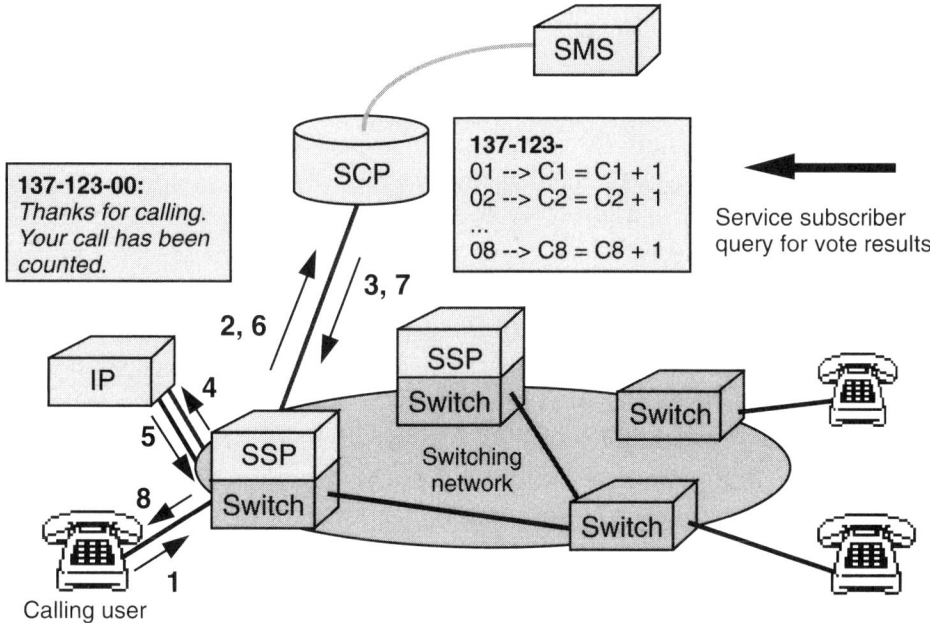

**Figure 3.25** Televoting service processing with originating user prompter.

the last digit dialed, the corresponding vote counter will be incremented. In addition, the SCP determines an appropriate user interaction device, i.e. an IP, which contains an announcements saying that the call has been counted successfully.

3. The SCP returns a routing number of an appropriate IP containing the televoting announcements to the SSP.
4. The SSP routes the call to the IP and instructs the IP to play the announcement to the user. The IP plays the desired announcement (e.g. 'Thanks for Calling. Your Call has been counted').
5. The IP informs the SSP that the announcement has been presented to the calling party.
6. The SSP relays this information to the SCP.
7. The SCP instructs the SSP to disconnect the IP and to clear the call. Note that in the basic televoting service processing no real connection to a subscriber installation will be established. Nevertheless, televoting implementations that connect the calling party to a subscriber installation after voting are possible.
8. The SSP instructs the switch to clear the call.

### 3.1.4.4 Card calling

Card calling services offer the possibility that a phone call can be charged to the account of the user who makes the call, instead of the to user corresponding to the calling or called line. Thus, the call can be established without charging any of the lines involved in the call. This service is also known as automatic alternative billing, account card calling, virtual card calling, or credit card calling service (section A.1). The distinction between all of these services depends on whether a real physical card is being used (which means that a card-reading terminal must be available) or a 'virtual card' is used (this means that the caller has to dial an account number).

All these services are based on the existence of a 'white list' managed by the service administrator. This list, residing in a database, contains an updated relation of subscribers to these services and related billing information. The database is administered by the management system that accepts service order information or receives updates through terminal access. The basic function of the IN consists in validating and checking account number and PIN (personal identification number) digits. The PIN is used for validation.

It should be emphasized that, in general, in this IN service the calling party, i.e. the service user, is the service subscriber, as he or she owns the calling card.

However, in some cases the call card may belong to the company for which the user works. In the next section we look at a virtual card calling service, which is defined by the automatic alternative billing service.

*Basic service functionality*

The automatic alternative billing service allows subscribers to call from any normal network access interface to any destination number and allows calls to be charged to an account specified by the account number. The line from which the call originates will not be charged. The account number and PIN are allocated to the service user by a service management procedure.

To invoke the service, the user must dial the service access code (which could be, for example, a Freephone number) followed by the destination number. The user will subsequently be asked for the account number and PIN. The account number and the PIN are validated; this necessitates the user dialing a rather long sequence of digits in order to make the call.

Thus, the basic service functionality is defined by the two service features 'authorization code' and 'originating user prompter' (section A.3):

- The *authorization code* service feature allows users to obtain special calling privileges depending on their authorization profile. Different sets of calling privileges can be assigned to different authorization codes, with codes able to be shared by different users. This feature was first introduced for virtual private networks (VPNs) to allow users within a VPN to override certain calling limitations of a VPN station.
- The *originating user prompter* service feature is described in the Freephone example.

*Optional service functionality*

In accordance with the table in section A.3, the following optional service features may be incorporated within a specific televoting implementation (for a description of all of these service features readers are referred to section A.2):

- abbreviated dialing
- call logging.

The first service feature allows subscribers to use their own set of short numbers (e.g. a two-digit code may represent a complete phone number), whereas the second one enables them to monitor service use.

*Service example*

In this section we consider a user who has an account card and wants to place a call from a phone that has no card-reading device. To do this, the user dials the service access code for using the automatic alternative billing service (e.g. '123-123') followed by the destination number (e.g. 030-25499-229). After gaining access to the network the user will be asked for the account code (e.g. '111') and PIN (e.g. '8888'). After account code and PIN validation the call will be established.

*Distributed service processing*

The call processing for the above scenario is as follows (Figure 3.26).

1. The user dials the service access code followed by the destination number.
2. The switch recognizes that the call is an IN call and the SSP sends an INAP query containing call information to the corresponding SCP. On receipt of the query the SCP starts the corresponding service logic program. The SCP determines an appropriate IP to query for the account code and PIN of the user for validation.

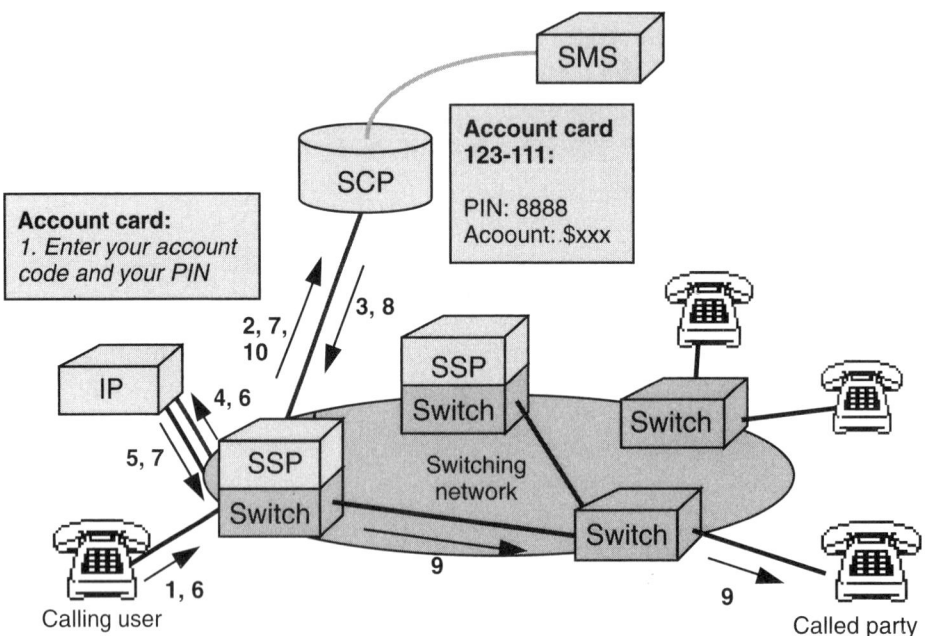

**Figure 3.26**  Automatic alternative billing service processing.

3. The SCP returns to the SSP a routing number of an appropriate IP and instructs the SSP to establish the connection to the IP.

4. The SSP routes the call to the IP and instructs the IP to start an appropriate dialog with the user.

5. The IP asks the user for the account code and PIN.

6. The user enters the account code and PIN. The IP collects the response digits.

7. The IP returns the information to the SSP. The SSP relays the account code and the PIN to the SCP. The SCP examines the account code and the PIN and checks that the account limit has not yet been exceeded. (Optionally the SCP may check that the user is allowed to place a call to the desired destination. However, this is only necessary in case of the terminating call screening service feature.)

8. The SCP instructs the SSP to disconnect the IP and to establish the connection to the destination number.

9. The SSP disconnects the IP and instructs the switch to establish a connection to the desired destination.

10. After the call is terminated, the SSP informs the SCP about the call charges. The SCP adds the charges to the subscriber's account.

## 3.2 IN capability set 2

IN capability set 2 (CS-2) is the second standardized stage of the IN as an architectural concept for the creation and provision of telecommunication services. The identified work areas of CS-2 standardization (ETSI Group NA6, 1993; ETSI ETS MI NA-60003; ITU Draft Recommendation Q.1221) are primarily addressing the limitations of CS-1, as follows.

- Enhanced IN service capabilities should be supported, such as mobility and broadband services.
- IN interworking should be supported to enable international service provision.
- IN management should be investigated based on common management standards.
- IN service creation should be studied in order to harmonize IN platform development.

The development of CS-2 standards is also based on the INCM, that is the development of an enhanced IN architecture begins again with the identification of envisioned service capabilities in the service plane. This enhanced service scope has effects on all lower planes of the INCM. In the GFP new SIBs have to be identified and the BCP SIB has to be enhanced. In the DFP the BCSM has to be appropriately enhanced. Additionally, the identified functional entities have to be enhanced as well as new entities defined, resulting in the definition of new IFs. In the physical plane the IN application protocol has to be enhanced to accommodate the new IFs. New physical entities may also have to be defined.

In the following section we will focus only on the major enhancements of CS-1, as CS-2 standards have not yet been finalized at the time of writing. Thus, the information given in this chapter may be subject of changes. However, the standards in Table 3.9 are currently under construction:

Table 3.9 ITU CS-2 recommendations

| Document no. | Title |
|---|---|
| Q.1221 | *Introduction to Intelligent Network Capability Set 2* |
| Q.1222 | *Service Plane for Intelligent Network CS-2* |
| Q.1223 | *Global Functional Plane for Intelligent Network CS-2* |
| Q.1224 | *Distributed Functional Plane for Intelligent Network CS-2* |
| Q.1225 | *Physical Plane for Intelligent Network CS-2* |
| Q.1228 | *Intelligent Network Interface Recommendations* |
| Q.1229 | *Intelligent Network Users Guide for Capability Set 2* |

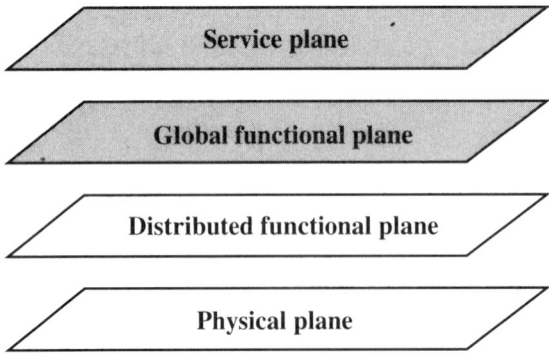

**Figure 3.27** CS-2 services are defined by the upper two INCM planes.

This section is structured like 3.1. In section 3.2.1 we consider briefly the envisioned enhancements of the upper two INCM planes (Figure 3.27), looking at CS-2 services, service features, and in particular the SIBs. In section 3.2.2 we examine the corresponding enhancements of the IN architecture, focusing primarily on the DFP.

## 3.2.1 CS-2 services

In this section we look at the enhancements of the INCM service plane and the GFP for IN capability set 2.

### 3.2.1.1 CS-2 service plane

According to the IN modeling methodology, the CS-2 services and service features are the starting point for the definition of CS-2 SIBs, the call-processing model, and the service control principles. In general, the CS-2 service and service features are based on CS-1 services and service features, with services dealing with certain aspects of mobile communications (including both personal mobility and terminal mobility) and broadband ISDN added. In addition, inter-network signaling aspects required for the provision of international IN services across multiple INs have been incorporated. Finally, the distinction between call control and connection control is another important aspect of the CS-2 services in the light of emerging multimedia and mobile communication services.

Instead of concrete services, the following 'service categories' have been identified for IN CS-2 (ETSI DTR NA-60902; ITU Draft Recommendation Q.1221 Annex B):

- *internet working services*, such as Internet Freephone, Internetwork Premium Rate, Internet Televoting, and Global virtual network services based on interconnected INs;
- *call party handling* services allowing to manage various parties' participation within a call;
- *mobility services* and features required for personal mobility (e.g. universal personal telecommunications, UPT), terminal mobility (e.g. support of cordless terminal mobility, CTM, in IN-structured networks), and the emerging third-generation mobile communication systems (e.g. universal mobile telecommunications system, UMTS);
- *broadband services* and *bearer services* (with lower priority), including connectionless and connection-oriented bearer services on demand;

- other services features outside the range of 'single ended' and/or 'single point of control' (e.g. conference calling and completion of call to busy subscriber) that were not fully addressed within CS-1 and some new individual telecommunication services.

In addition to these service categories, ITU also plans 'service management services' and 'service creation services' (ITU Draft Recommendation Q.1221). A list of envisioned CS-2 service features is given in Appendix B of this book.

It must be recognized that the definition of new services (and service features) results in corresponding enhancements of the lower planes of the INCM, such as new SIB definitions, call model enhancements, definition of new functional entities, and enhancements of the IN application protocol.

### 3.2.1.2 CS-2 global functional plane

The enhanced scope of CS-2 service features must be supported by corresponding concepts in the GFP. Thus, the following enhancements have been made to the CS-1 GFP (ETSI DTR NA-60801; ITU Draft Recommendation Q.1223).

- Flexible granularity of SIBs has been added, in which 'composition' and 'decomposition' of SIBs will be supported. Composition means that SIBs can be made out of smaller SIBs, forming *high-level SIBs* (HLSIBs). High-level SIBs are, like normal SIBs, a reusable part of a service feature. HLSIBs can be composed out of other high-level SIBs and normal SIBs. However, a particular HLSIB cannot be used as a component within the same HLSIB (no recursion). On the other hand, SIB decomposition allows the partitioning of the granularity of HLSIBs into smaller blocks that can be reused.
- The concept of parallelism allows the global service logic (GSL) to be decomposed into several SIB chains that may perform service processing activities in parallel. Thus, the concept of *service processes* has been introduced to allow for parallel execution, within each service process the service logic being executed sequentially. Hence, the GSL is formed by several SIB chains, namely service processes, that execute different activities related to the same service. Therefore appropriate means for service process creation and interservice process synchronization and communication have been defined. Synchronization and communication between two service processes can be achieved via *points of synchronization* (POS).

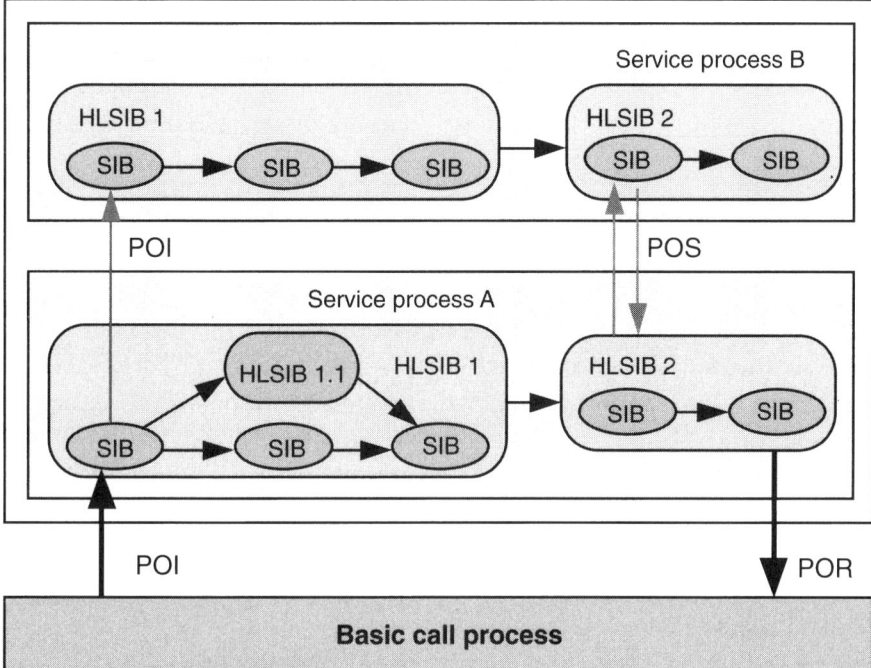

**Figure 3.28**  CS-2 global functional plane modeling concepts.

- Interworking between different service processes belonging to different 'domains' should be visible at the GFP level.
- Additional SIBs are defined to support advanced call party handling, such as is required for manipulation of multiple connections for multi-party calls.

A consequence of the concept of decomposition of the GSL into several service processes is that corresponding communication mechanisms have to be provided. Communication between two service processes can be achieved via POS. A POS is a functional interface between service logic of two service processes over which asynchronous communication is initiated. A particular SIB in the sending process has the ability to send a synchronize signal to a different service process that is executed in parallel. The receiving service process may use a particular SIB to wait until the synchronize signal arrives.

Figure 3.28 illustrates the relationships between these GFP enhancements. The GSL can be achieved by two service processes executed in parallel. After invocation of service process A, it spawns a new service process B in HLSIB 1. In HLSIB 2, service process A synchronizes with service process B, before it returns control to the BCP.

All these considerations have resulted in the enhancement of existing CS-1 SIBs (section 3.1.1.2) e.g. 'user interaction', compare, and the definition of the following additional SIBs for CS-2.

- *Basic Call Unrelated Process* (BCUP) is a specialized SIB which enables non-call associated interactions.
- *Attach* establishes a connection to a remote resource.
- *Detach* removes an indicated call party from a call.
- *Join* establishes a speech path between two indicated call parties.
- *Service filter* filters the number of calls related to IN-provided service features.
- *Split* breaks the continuity of the speech path of the indicated call party.
- *Initiate service process* starts a parallel service process instance.
- *Send* sends a signal (conveyed with interprocess data) to another service process that is executing in parallel.
- *Wait* waits for the synchronization signal (with interprocess data attached) sent by another service process.
- *End* indicates the normal end of an executing service process, or part of a service process in case of multiple threads.

Note that the last four SIBs are required to support the enhancements of the GFP modeling.

A good overview of the CS-2 GFP enhancements is also given in Hu (1995).

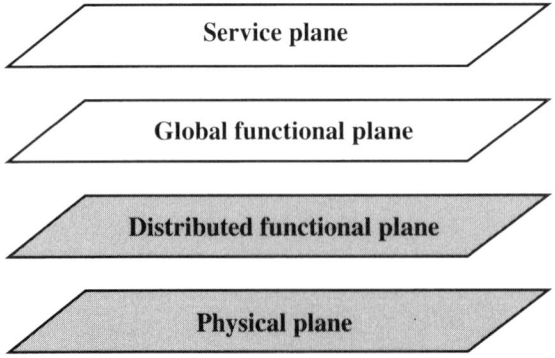

**Figure 3.29** The CS-2 IN architecture is defined by the lower two INCM planes.

## 3.2.2 CS-2 architecture

The envisioned CS-2 services require enhancements of the CS-1 call-processing model and the CS-1 functional model. For this reason this section focuses on the CS-2 IN architecture defined by the two lower planes of the INCM (Figure 3.29).

### 3.2.2.1 CS-2 distributed functional plane

For CS-2, the CS-1 DFP has been enhanced with respect to the following points (EST DTR NA-60401; ITU Draft Recommendation Q.1224).

- The impact of mobile communication systems, particularly the *universal mobile telecommunications system* (UMTS) (ETSI DTR SMG-50301), on the IN functional model requires the support of 'location management' (used for terminal mobility provision and comprising functions for terminal attachment/detachment, location registration and location updates) service features. As these service features require the invocation of IN capabilities (i.e. access to the SCF) outside the context of a call, additional functional entities have been defined.

- The adoption of international management standards, i.e. the *telecommunications management network* (TMN) standards (ITU Recommendation M.3010), has resulted in the definition of a separate IN management functional model, identifying the relevant TMN functional entities.

- Interworking between IN-structured networks is supported by specifying SCF–SCF and SDF–SDF interfaces.

- The impact of broadband ISDN on the IN functional model requires that several connection instances have to be controlled via the SCF–SSF interface. Thus, a separation of *call control* (CC) and *bearer control* (BC) within the switch is the subject of further research that will result in the definition of additional functional entities (see also section 4.3 for further details).

The traditional CS-1 functional architecture is therefore, separated in CS-2 into two complementary DFP models (ETSI DTRNA-60401), namely:

- the IN *service execution functional model*, which addresses the execution aspects of an IN service, representing an enhancement of the CS-1 DFP; and

- the IN *management functional model*, which addresses the management aspect of the IN DFP, based on standardized management architectures, i.e. TMN.

*Service execution functional model*

The CS-2 service execution functional model is based on the DFP defined within CS-1. However, Figure 3.30 shows that three basic changes from the CS-1 functional architecture can be recognized.

1. All management-related functional entities, i.e. the service management agent function (SMAF), service management function (SMF), and service creation environment function (SCEF), have been removed, as a separate IN management functional model has been defined (see next section for details).

2. The requirement to support interworking between INs and between INs and private networks has resulted in the definition of two new interfaces, i.e. the SCF–SCF interface and the SDF–SDF interface. Thus, interactions between different SCFs and between different SDFs can be supported. It must be stressed that these entities could belong to different network domains, which allows for IN interworking. IN interworking issues will be addressed in more detail below (pp. 104–107).

3. In order to support mobility applications and the related non-call-associated services features, such as location management service features, which require the invocation of IN capabilities outside the context of a call, two new functional entities have been defined. These are the *service control user access function* (SCUAF) and the

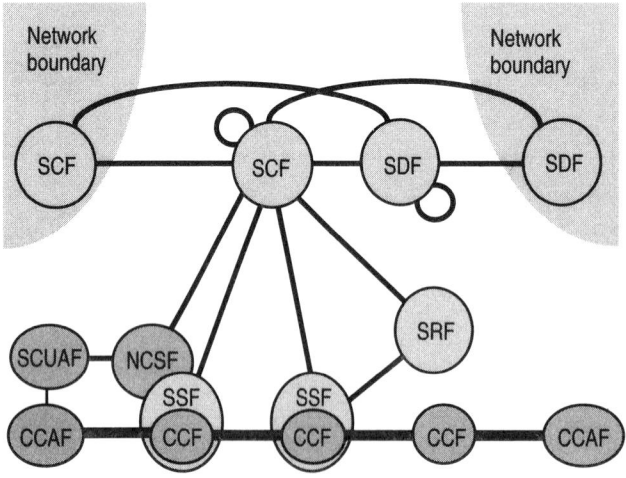

**Figure 3.30**   Service execution functional model for CS-2.

*non-call service function* (NCSF). The SCUAF is defined in order to establish, maintain, modify, and release, as required, an instance of a service. It provides access for users/terminals and accesses the service invocation capabilities of the NCSF, using service requests (e.g. location registration, attach, etc.) for invocation of non-call associated services. The NCSF, which is associated with the connection control function and the service switching function, provides a set of functions required for access and interaction between the user and the service control function for non-call associated services. The NCSF may be renamed into Call Unrelated Service Function, CUSF. Consequently, new interfaces have been defined for the new SCUAF and NCSF functional entities as depicted in Figure 3.30. One important application in this context is 'cordless terminal mobility (CTM)' phase 1, which is currently defined by ETSI Draft DE NA-10039 and ETSI DTR NA-61302. (See also section 4.2, which addresses the relationship between IN and mobile communications.)

It is important to note that the ITU, in its CS-2 DFP, follows more closely the requirements for supporting terminal mobility resulting from the requirements raised by third-generation mobile communication systems, such as the future public land mobile telecommunications system (FPLMTS)/UMTS. (The FPLMTS was developed by ITU. UMTS was developed by the ETSI and is closely related to FPLMTS. Section 4.2 considers these systems in more detail.) Hence, the architecture enhancements concentrate on the support of radio systems. Thus, the ITU has defined a *radio access control function* (RACF) and a corresponding *radio link function* (RLF). The RACF provides functionality for the processing and control within the environment of a given radio system. It provides support for non-call-associated service features in the wireless environment (e.g. terminal registration). It also provides radio-specific processing and control support for service features requiring the handling of a radio link that is related to a call (e.g. radio link set-up). The RLF assists in providing access to users/terminals. It is the interface between the user/terminal and the network call-related radio link functions as well as the interface to the user/terminal and the network non-call-associated functions. More information on these functions can be found in ITU Draft Recommendation Q.1224.

*Management functional model*

The definition of IN management concepts is of pivotal importance for the success of the IN, as the rapid introduction of new services enabled by the IN architecture concept requires the corresponding provision of IN service and network management capabilities. These management capabilities are related to the installation and maintenance of IN service software in the IN architecture, as well as the management of the IN network elements themselves. CS-1 has only defined the need for a service management function, with the specification of common management interfaces considered to be outside the scope of standardization. This approach greatly hinders the concept of vendor independence, as each vendor will develop its own management interface. This also makes the interworking of different IN platforms, e.g. for providing international IN services, a particularly complex issue, as different IN management systems with proprietary interfaces have to be interconnected. An *IN* management functional model has therefore been developed.

This management functional model should enable both the effective management of the network infrastructure required to support IN-based services and the efficient management of IN-based services, including service creation, service deployment, service provision, and service maintenance.

The definition of the IN management functional model has been primarily influenced by the emerging TMN (ITU Recommendation M.3010) standards. (For an introduction to TMN, readers are referred to section 4.1, which addresses the relationship between IN and TMN.) In other words, CS-2 proposes a *TMN-based management solution* for IN, as TMN provides the

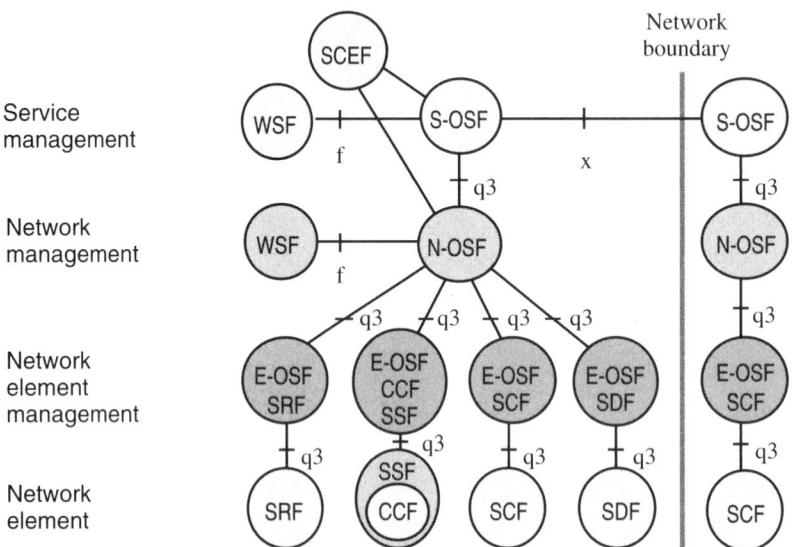

**Figure 3.31** IN management functional model for CS-2.

capabilities to manage telecommunications services, networks, and network elements in a uniform way. This step is straightforward, as TMN is the general management framework for the overall telecommunications environments, of which IN is only one part (Magedanz *et al.*, 1993; Magedanz 1994), that is a common management philosophy is adopted for the management of the IN and other networks (e.g. ISDN) services and equipment allowing better economies through the use of common techniques (e.g. reuse of management software for different applications).

Based on the ETSI document *Study on the Relationship of IN and TMN* (NA-43308) and results from various IN/TMN-related research projects, the IN management functional model (Figure 3.31) adopts the TMN functional hierarchy approach, which distinguishes four layers of management functionality, namely *business, service, network*, and *network element* management. Accordingly, the IN functional entities are regarded as network elements to be managed, i.e. they represent the 'managed system'.

The management functional model replaces the management-related CS-1 functional entities, i.e. the service agent management function (SMAF) and the service management function (SMF), with corresponding TMN functional entities. A *workstation function* (WSF) provides the user with access to an *operations system function* (OSF), which supplies the required management functionality, such as service installation, maintenance, and accounting management, i.e. the WSF replaces the SMAF and the OSF replaces the SMF. In the TMN functional hierarchy, a layering OSFs has been adopted. The *service-OSF* (S-OSF) is responsible for IN service management. The *network*-OSF (N-OSF) is responsible for managing the IN-structured network as a complete entity, whereas the multiple network *element-OSF (E-OSF)* provides the management capabilities for specific IN network elements (i.e. a specific E-OSF manages an SSF, another E-OSF manages an SCF, etc.).

Correspondingly, the reference points between the OSF(s) and the IN functional entities are modeled as TMN 'q3' reference points within one IN domain. This means that IN management information has to be modeled in accordance with OSI management principles (ISO-10165-1). In addition, the IN management model also considers IN interworking, which requires an interconnection of the related S-OSFs via TMN '*x*' reference points. These reference points will become corresponding Q3 and X interfaces in the physical architecture, based on OSI's common management information service/protocol (CMIS/P) (ISO-9595; ISO-9596-1). Additionally, a WSF interacts via an 'f' reference point with an OSF. The relationship between the service creation environment function (SCEF) and OSFs for management service creation is the subject of further study.

Section 4.1 will provide further information on the subject of IN management and IN/TMN integration.

*IN interworking*

In the light of progressing deregulation of the telecommunications environment and the need for international IN service offerings, as identified in the service plane, interconnecting different INs becomes an important issue (IEEE, 1993b). The principal aim is to provide a common 'look and feel' to IN services in different countries. Currently the services offered by different countries have functionalities based on distinct IN platforms. To achieve international services, the service calls will always be routed (via a corresponding bearer connection) to the IN platform implementing the service. This may be particularly inefficient in the case of number translation-type services, when both calling party and the desired called party are located in a remote network.

Thus, the provision of international IN services, e.g. an Internetwork Freephone service with a common service access code from any telephone in Europe, requires the interworking of different (national) IN platforms (Covaci *et al.*, 1995) to avoid the establishment of superfluous bearer connections. CS-2 therefore identifies two areas of IN interworking:

- interworking between IN-structured networks (ETSI DTRNA-60301); and
- interworking between private networks (IN-structured and non-IN structured) and public IN-structured networks (ETSI DTRNA-60302).

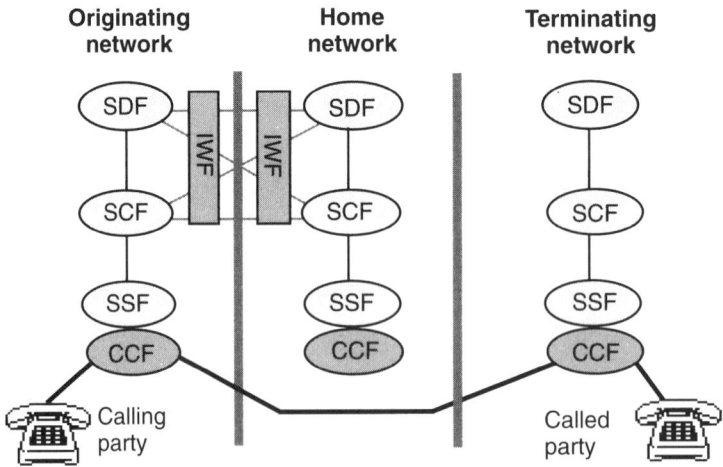

**Figure 3.32** IN interworking scenario.

The EURESCOM project P230, 'Enabling Pan-European Services based on Interconnected IN Platforms', has contributed substantially to these documents. The second aspect is important to support IN service provision also in cooperation with legacy network platforms. Below we briefly address the first point in more detail. International IN services require interworking of different national IN platforms, based on standard interfaces. Otherwise, in the case of $n$ different IN platforms $n * (n–1)$ specific interworking interfaces/protocols have to be developed to support complete interworking. In the case of a standardized interface only, each platform has to develop a corresponding interface. For IN interworking in general three logical types of network can be identified, as depicted in Figure 3.32.

- *Originating network:* provides the calling user with access to the network and to the services. In order to provide this service access, some specific service logic and data is required in the originating network.
- *Home network:* where the service subscriber is administered and where the service logic and data, particularly the customer data (customer profile), resides.
- *Terminating network:* provides access to the called user.

Note that these network categories are only logical ones, with one physical network potentially having multiple 'roles'. The home network could also act as the originating and/or terminating network and the originating network could also be the terminating network. Note that the home network is not necessarily involved in the bearer connection between the originating and terminating networks (recently replaced by the terms invoking and responding respectively).

Let us consider, for example, a pan-European Freephone service, offered in France, Germany, and Italy. The subscriber may be a major corporation such as Lufthansa, with its main office in Germany and several branches in France and Italy. Hence the subscriber data, i.e. the customer profile containing all relevant call-handling attributes, will be kept in the German IN, acting as the home network. Users could access the Freephone service in all three countries by dialing a common service number. Each country's IN acts as the originating network. Depending on the routing policy defined in the customer profile, each country's network could also act as the terminating network (e.g. during normal working hours each call would be routed to the next Lufthansa office).

Consider a case in which all calls received after 18.00 hours should be routed to one specific Lufthansa office in Italy, the Italian network thus acting the terminating network for all calls originated after 18.00 hours. When someone in France dials the Freephone number, the French IN acts as the originating network, which then has to contact the German IN to obtain the appropriate destination number. After receiving the required routing information, the French network could set up a direct bearer connection to the Italian network, i.e. the German (telephone) bearer network is not involved in the resulting call connection. As the service subscriber (i.e. Lufthansa) is charged for Freephone calls, the French IN would transfer the related call charges after call completion to the German network.

In accordance with the example interworking scenario, specific service logic and data has to be distributed in the home and the originating networks. In most cases, i.e. for most service features, the terminating network does not require any IN service logic and data. Exceptions are the 'queuing' and 'call limiter' service features, in which cases status information of the terminating line is required by the service logic in order to perform appropriate call handling. In contrast to the home network, which contains the main service logic and data and in particular the customer profile, the originating network needs specific service triggers, logic, and data in the service switching function, the service control function, and in some cases in the service data function to request the required call handling information from the home network.

Possible interworking scenarios identified are:

- SCF–SCF interconnection, that is two service logic programs interact directly, with access to subscriber data in the SDF only possible by the local SCF;
- SDF–SDF interconnection, that is an SDF is able to access another network's SDF without the support of the SCF (i.e. service logic program) [this can be compared to the implementation of the X.500 directory service, which is achieved by distributed interacting directory system agents (DSAs) (ITU Recommendation X.500)];
- SCF–SDF interconnection, that is an SCF may access another network's SDF directly.

A newly defined *interworking function* (IWF) provides the specific interworking capabilities, such as security functionality and protocol/data conversion capabilities. More details on IN interworking can be found in ETSI DTRNA-60301.

At the current stage of deregulation, it seems likely that the first scenario for IN interworking is the most likely one. The reason for this is that IN service providers do not like to offer direct access to subscriber data to third parties. Implementing this access via the SCF provides them with greater access control.

### 3.2.2.2 CS-2 physical plane

At the time of writing only limited information was available regarding the CS-2 physical plane (ITU Draft Recommendation .1225). However, it can be stated that the mapping of CS-2 functional elements, identified in the service execution functional model, onto physical entities is analogous to CS-1 mapping. In addition, the two newly defined functional entities, i.e. the service control user agent function (SCUAF) and the non-call service function (NCSF) have to be allocated to a corresponding physical entity. It seems most likely that the NCSF will be co-located with the service switching function within the service switching point or within a network access point. The SCUAF will be probably co-located with the connection control agent function within the service switching point or within a network access point. Optionally, a service node or a service switching and control point may host these two functions (ITU Recommendation Q.1225).

Looking at the INAP supporting the interactions between the IN physical entities, it must be recognized that an enhanced version is required to support all the enhancements made in the DFP, in other words new application service elements, comprising appropriate INAP operations, have to be defined in order to support:

- SCF–SCF interactions;
- SDF–SDF interactions (it seems most likely that these will be achieved by means of the X.500 directory system protocol; ITU Recommendation X.500);
- NCSF–SCF interactions; and
- SCUAF and NCSF interactions.

A new version of INAP is due at the end of 1996 (ITU Recommendation Q.1228).

For completeness it should be mentioned that the functional entities identified in the management functional model will be implemented within the TMN physical architecture (see section 4.1 for details).

## 3.3 Future capability sets

The content of future IN capability sets is difficult to predict, as the IN commonly considered as the ultimate service control architecture is currently

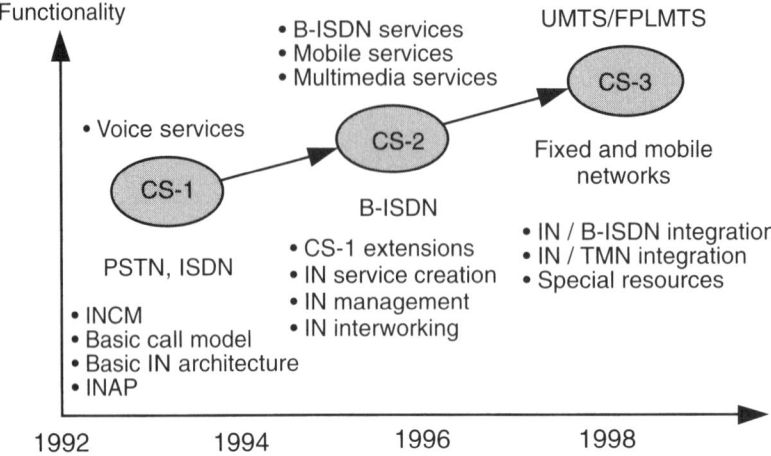

**Figure 3.33** IN capability set evolution.

influenced by many other telecommunication evolution trends (Figure 3.33). Consideration of CS-2 reveals that service aspects, such as mobility, broadband, and multimedia (Hetz, 1992; ITU Study Group XI, 1992; ETSI DTR NA-60107 ITU Recommendation Q.1221), are regarded as basic attributes of future IN services, and these must be supported by corresponding IN architectural enhancements. Considering the time required for CS-1 completion, it can be assumed that these issues cannot be solved completely within the given time frame for CS-2 finalization (i.e. spring 1997). Thus, work related to CS-3 will include the completion of these challenging aspects, and in particular address issues such as full IN/TMN integration, full IN/B-ISDN integration, and full support for mobile/personal communications systems (i.e. FPLMTS/UMTS).

In this context the relationship between the aspects of 'signaling', 'service control', and 'service management' becomes a challenging issue. Current approaches to IN/TMN integration (Berndt *et al.*, 1995; Magedanz, 1995), and IN/B-ISDN integration (Bretecher & Vilian, 1995) indicate that significant changes in call modeling will affect call models, signaling protocols, and in particular the IN architecture in general. In Chapter 4 we will address the relationships between IN and TMN, mobile communications, and B-ISDN in more detail.

Nevertheless, it has to be recognized that, in parallel with the international IN standardization, other initiatives involve the study of the long-term evolution of IN, taking into account the recent advances in the field of distributed computing. Here asynchronous communication services (e.g. electronic mail) as well as multimedia applications requiring quite different applications support are gaining momentum and thus can be seen as basic drivers toward new information networking architectures. This trend and the convergence of computing and telecommunications and the increasing influence of the object-oriented paradigm in the telecommunications environment in general will probably have additional effects on future IN capability sets, and these will be addressed in more detail in Chapter 5.

# 3.4 Other IN standards – AIN

In the previous sections of this chapter we have concentrated on the international IN standards defined by the ITU and the ETSI. However, it is important to note that intelligent networks have been studied in various organizations throughout the world. The other big player in the context of IN 'standardization' is Bellcore, with its *advanced intelligent network* (AIN) program in the US. (AIN specifications can be obtained from Bellcore Customer Relations, 8 Corporate Place, Room 3A-184, Piscataway, New Jersey 08854-4156, USA, Tel +1 908 699 5800, Fax +1 908 336 2559.) In this section we do not aim to provide a comprehensive description of AIN concepts and specifications, as the focus of this book is the international IN standards. Rather, we want to provide an overview of the commonalities and differences of ITU IN standards and AIN specifications.

AIN can be seen as a parallel IN 'standardization' activity in North America, mainly driven by Bellcore. AIN documents are Bellcore specifications, referred to as 'generic requirements', and do not represent real international standards. However, these specifications are considered to be standards in North America as they are used by the Bellcore client companies to implement AIN services. The AIN generic requirements cover the following areas:

- switching system, i.e. service switching point (SSP) requirements;
- service control point (SCP) requirements;
- switching system (SSP)–service control point (SCP) application protocol interface generic requirements;
- operations system (OS)–service control point (SCP) interface generic requirements;
- intelligent peripheral requirements.

## 3.4.1 AIN evolution

Work on AIN started in 1989 as a follow-up to the IN-1+ developments and is performed jointly by Bellcore and the Multi-Vendor Interaction Forum

(MVIF), which was formed in the US at the end of 1987. The AIN program was developed to address IN evolution in the local exchange carrier networks of the Bellcore client companies in the US, where AIN is considered to provide an architectural solution for the 1992–98 time frame.

Like the ITU approach, AIN is developed in evolutionary steps, so-called *'releases'*, which can be compared to the ITU capability sets, although the capabilities and timing of these increments do not necessarily coincide with the capabilities and timing of the ITU capability sets. This means that there exists no one-to-one mapping between specific AIN releases and ITU capability sets.

Bellcore's current target is referred to as AIN release 1, which is viewed as the target AIN architecture for the delivery of circuit-switched voice/data services. However, it must be pointed out that earlier Bellcore publications mentioned AIN release 2, with a much earlier target date for AIN release 1, but AIN 2 was abandoned by postponing AIN release 1 and introducing the subsets AIN 0.1 and AIN 0.2. Like the ITU approach of working concurrently on a given IN capability set and on refinements to the target IN architecture, Bellcore is working concurrently on the definition of AIN releases leading to AIN release 1. Thus, the generic requirements are packaged as individual documents based on release number and technical area.

- The first of these releases, *AIN release 0.0,* was finalized 1992; this was not a pure Bellcore specification but a common denominator for different regional Bell operating companies' (RBOCs) IN implementations relying on IN-1+ (Bellcore, 1988).
- The next release, *AIN release 0.1* (Bellcore, 1992a,b), was targeted for 1993–95 and was defined by Bellcore using MVIF results. AIN 0.1 provides a better separation between the logical and physical architecture than IN-2, allowing for easy integration of already existing AIN 0.0 implementations.
- *AIN release 0.2* (Bellcore, 1993d–f), which should continue the progress of AIN release 0.1 and carry AIN even closer to its objectives, is targeted for 1995–97.
- To guide this evolutionary AIN approach, Bellcore issued generic requirements for the initial view of AIN release 1 in 1989–90 (Bellcore, 1989, 1990), followed by updates in 1993 (Bellcore, 1993a,b) and 1994 (Bellcore, 1994a–c).

Putting this together the major Bellcore AIN documents in Table 3.10 are available (the table is not intended to be complete).

Table 3.10 Bellcore AIN documents

| Release | Document no. | Document title |
|---------|--------------|----------------|
| AIN 0.1 | TR-NWT-001284 | *AIN 0.1 Switching System Generic Requirements* |
| AIN 0.1 | TR-NWT-001285 | *AIN 0.1 Switching System–Service Control Point Application Protocol Interface Generic Requirements* |
| AIN 0.2 | GR-1298-CORE | *AIN 0.2 SSP Requirements,* Issue 1 |
| AIN 0.2 | GR-1299-CORE | *AIN 0.2 SSP–SCP /Adjunct Application Protocol Interface Generic Requirements,* Issue 1 |
| AIN 0.2 | GR-1229-CORE | AIN 0.2 SSP–IP Interface Generic Requirements, Issue 1 |
| AIN 1 | GR-1280-CORE | *AIN SCP Generic Requirements* |
| AIN 1 | GR-1286-CORE | *AIN OS–SCP Interface Generic Requirements,* Issue 1 |
| AIN 1 | GR-1298-CORE | *AIN SSP Requirements,* Issue 2 |
| AIN 1 | GR-1299-CORE | *AIN SSP–SCP /Adjunct Application Protocol Interface Generic Requirements,* Issue 2 |

It can be seen that Bellcore's work has focused primarily on the lower two planes of the INCM, namely the DFP and the physical plane. As with ITU, Bellcore used both a 'bottom-up' and 'top-down' approach to define AIN release 1 target and interim releases. The 'bottom-up' approach was used primarily to establish the AIN release 1 target, and the 'top-down' services approach was used and continues to be used to identify the appropriate AIN capabilities needed for a particular release. Thus, the methodologies used by ITU and Bellcore are aligned to a large extent. However, it has to be stressed that there is no notion of the INCM within the AIN specifications, nor does any AIN specification specify AIN services, service features, and SIBs explicitly.

## 3.4.2 AIN services

AIN 0.1 and AIN 0.2 are closely aligned with ITU capability set 1 in terms of supporting circuit-switched voice/data services, with an emphasis on flexible routing, flexible charging, and flexible user interaction for two-party calls. As with CS-1 services, AIN 0.1 and AIN 0.2 services are considered to be single ended, and are expected to interwork with existing switch-based services. AIN release 1 extends to the single-ended service concept to address service capabilities that apply during the active phase of a call and to multiparty calls.

Additionally, it has to be stressed that AIN considers already the problem of feature interactions very carefully. That is the AIN releases specify switch-based feature behavior in the presence of AIN.

## 3.4.3 AIN architecture and interfaces

AIN 0.1 and AIN 0.2 are subsets of AIN release 1, which in turn is a subset of the long-term evolution of the AIN architecture. As with CS-1, AIN 0.1 focuses on service processing requirements and minimum network interworking requirements for AIN 0.1 services. However, AIN 0.1 goes beyond CS-1 by addressing also service management and network management requirements. AIN 0.2 and AIN release 1 address additional requirements in all of these areas, but like CS-1 they do not address requirements for service creation.

From a service processing perspective, the AIN release 1 and ITU CS-1 functional architectures are aligned. AIN identifies the following functional entities: network access (corresponds to the CCAF), service switching (corresponds to the CCF/SSF), service logic and control (corresponds to the SCF), information management (corresponds to the SDF), and service assistance (corresponds to the SRF). These functional entities have since been allocated to physical entities for AIN, which are the focus of the interim AIN releases, i.e. AIN 0.1 and AIN 0.2. As such, functional entities are not explicitly addressed in AIN 0.1 and AIN 0.2.

As with CS-1, call modeling is used in AIN and is the foundation for the distributed architecture. This call modeling includes the basic call model (aligned with the CS-1 BCSM), the connection view (aligned with the CS-1 switching state model), and feature interaction mechanisms. AIN 0.1 and AIN 0.2 only address the basic call model, which has only 14 detection points instead of 18 as in the CS-2 BCSM. However, the AIN release 1 basic call model should be aligned with the ITU long-term IN BCSM.

The AIN release 1 physical architecture is identical to the CS-1 physical architecture, with the following exceptions: AIN release 1 does not support direct interfaces between the service control point and the intelligent peripheral or between the adjunct and the intelligent peripheral; AIN release 1 treats the intelligent peripheral and the service node identically in terms of the interface to the service switching point; and AIN release 1 supports generic interfaces between operations systems (i.e. service management systems in CS-1 terminology) and each physical entity.

AIN 0.1 and AIN 0.2 are subsets of the AIN release 1 physical architecture. In particular, AIN 0.1 includes the service switching point, the service control point, and interfaces between the service switching point and the service control point, between service control points, between service control points and service data points, and between the operations systems and the service switching points and service control points. AIN 0.2 adds an interface between the service switching point and the intelligent peripheral.

As with CS-1, the focus of standardization of AIN interfaces is on the application layer protocol, referred to as *AIN application protocol* (AINAP). It defines the set of conventions to manage the communications between applications to provide services in an AIN architecture. AINAP defines the application layer protocol operations transferred between physical entities and protocol procedures performed at each entity. The protocol operations and associated parameters and errors are specified using ASN 0.1. AINAP is a user of the ANSI transaction capabilities application part (TCAP) of the CCS7 signaling network and is optimized for the exchange networks of the Bellcore client companies in the US. AINAP is not fully aligned with CS-1 INAP in terms of protocol architecture and encoding. In general, INAP is more granular and modular than AINAP in terms of the scope of each operation and the grouping of operations into application service elements. However, AINAP and INAP are aligned for common ISDN user part and digital signaling system 1 parameters.

## 3.4.4 AIN summary

Looking at the current state of IN standardization, one may basically compare AIN release 0.2 (AIN-0.2) to ITU's revised version of capability set 1 (CS-1R). In general, it can be stated that AIN focuses on the lower two planes of the INCM and thus provides a greater level of detail than the ITU IN standards, which have a more broader scope. Thus, AIN allows for better interoperability of AIN-conformant multivendor products at the moment. Figure 3.34 provides a mapping of AIN releases to ITU capability sets.

In terms of the services that need to be supported and the need to be defined to achieve these services, the standards are quite similar. In addition, the

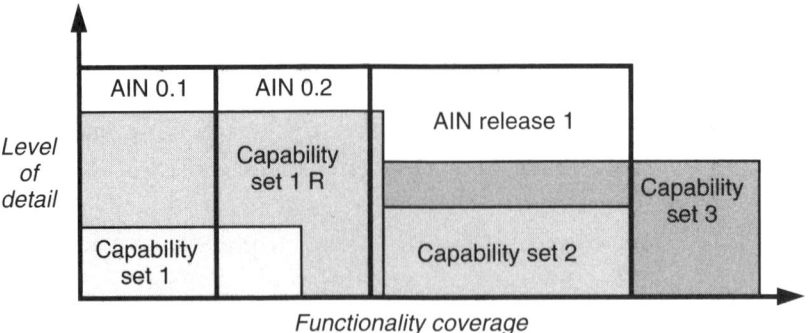

**Figure 3.34** Relationship between AIN releases and ITU capability sets.

defined network elements are the same. However, the basic differences between the two standards is use of slightly differing call models, resulting in minor differences in the information flows and the corresponding use of the TCAP portion of the CCS7 network. However, it is assumed that the call models of CS-2 and an enhanced version of AIN 0.2, referred to as AIN release 0.2+ are nearing strict convergence (Garrahan *et al.*, 1993; Turner, 1995). AIN 0.2+ is seen as the next step beyond AIN 0.2. This should include mid-call triggers as well as call party handling capabilities.

However, there is currently a common opinion that, depending on the progress of the ITU IN standardization of CS-2 and CS-3, potential AIN-1 successors may be replaced by international IN standards in the long term. In summary, it can be said that AIN releases and ITU capability sets are functionally converging with increasing speed.

In the remainder of this book we will no longer distinguish between ITU's IN and Bellcore's AIN, as we consider them to be mostly equivalent, i.e. the relationships between IN and other telecommunications systems and IN evolution are the same for both IN and AIN.

## 3.5 IN deployment and products

As the principal aim of this book is to provide a tutorial on IN, it is not our intention to provide a comprehensive overview of IN deployment and available IN platform products, as this information will soon be out of date. Nevertheless, we want to provide interested readers with some pointers to existing IN platforms.

Today INs are deployed all around the world. However, a leading position in IN deployment is held by North America (Russo *et al.*, 1993), Europe (Cancer *et al.*, 1993), and Japan (Suzuki, 1993). Many IN services are already offered on a national basis in some countries based on proprietary IN platforms. However, with the increasing globalization of national economies there is an emerging demand for international IN service offers, requiring the interconnection of national IN platforms. First, ITU standard-based IN platforms are likely to appear in the 1995–96 time frame in Europe, whereas AIN is still the common basis for IN deployment in the US and probably will remain so until the end of the decade. Looking at the first deployed IN services one may find in most countries Freephone/universal access number, televoting, premium rate, card calling services, and VPNs. Additionally, personal communication services are emerging. These services currently have the greatest market potential.

IN platform vendors include all the traditional switching equipment manufacturers, such AT&T, Siemens (Carl *et al.*, 1994), Alcatel (Drignath *et al.*, 1994), Northern Telecom, Ericsson, Nokia, etc., but the major computing vendors, such as IBM, Hewlett Packard, and Digital, are also enlarging their presence in the field, as INs can be seen to be integrators of telecommunications and computing.

For more detailed information on the issues of IN deployment and IN products interested readers are advised to consult the proceedings of three major international conferences, which provide probably the best overview in these fields. These are:

1. the annual IEEE Intelligent Network Workshop (INWS, 1994, 1996);
2. the biannual International Conference on Intelligent Networks (ICIN) (ICIN, 1992, 1994, which started in 1989 and is usually held in Bordeaux, France;
3. the first, second, third, fourth and fifth international telecommunications information networking architecture (TINA) workshops.

# 4 IN relations to other telecommunication systems

In the course of presenting the evolution of IN standards it has been mentioned that there are relationships between IN and other important telecommunication systems and that IN evolution will probably be affected by these systems. In this chapter we take a closer look at these relationships. Basically three major areas can be identified:

1. IN and telecommunications management network;
2. IN and mobile/personal telecommunications systems;
3. IN and broadband ISDN; and
4. IN relations to other telecommunications systems.

## 4.1 IN and telecommunications management network (TMN)

As outlined in this book, IN is considered to be the basic 'network' architecture for the implementation of sophisticated telecommunication services in the near future. The *telecommunications management network* (TMN) (ITU Recommendation M.3010) provides the worldwide accepted framework for an unified management of all types of telecommunication services and the underlying networks and their network elements (IEEE, 1995). TMN provides the basis for the uniform modeling of management services, management information and related management interfaces.

In this section we give a brief overview of the TMN standard, because we believe that a knowledge of TMN concepts is essential to understand the close relationships between IN and TMN and the development of management solutions for INs, such as the CS-2 management functional model described in section 3.2.2.

## 4.1.1 A brief overview of TMN

TMN standardization was started by ITU Study Group IV in 1985 and was first defined 1988 by the blue books as Recommendation M.30. Since publication of this recommendation several revisions have been proposed, resulting in November 1991 in a change in the number of the recommendation to M.3010.

In contrast to the early days of management standardization, when ITU's TMN was quite different from ISO's OSI systems management (ITU Recommendation X.700), which focuses on the management of open systems and the hosted services and applications in the OSI environment, the views of these concepts of the two bodies have converged, such that a coherent standards framework can be expected within the next few years. In some key areas for the future TMN, OSI management is ahead of ITU, wherein most of these areas ITU predicates on available ISO documents, adopting them in the X.700 series of ITU recommendations. This is especially true for ISO's definition of a common *structure of management information* (SMI) (ISO/IEC/IS 10165-1), based on an object-oriented approach defining management information in terms of *managed objects* (MOs), and the appropriate interface and protocol [common management information service CHIS/common management information protocol (CMIP); ISO/IEC/IS 9595, ISO/IEC/IS 9596-1] for the basic exchange of such management information.

The TMN framework defines a well-agreed network management architecture that provides a minimum but necessary platform for providing interoperability between different management systems and network components in the short term and a basis of an integrated management system for all kinds of telecommunication networks and services in the long term. A TMN (note that the term 'TMN' is used as a synonym for the ITU M.*xxxx* series of recommendations or denotes an existing management network belonging to one legal authority) is conceptually a separate network that interfaces with a telecommunication network (which it manages) at several points to receive information from it and to control its operations. A TMN may use parts of the telecommunication network to provide its own communication.

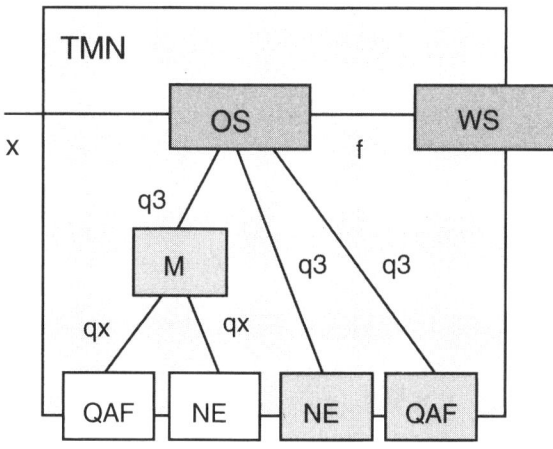

**Figure 4.1** Functional TMN architecture.

## 4.1.2 TMN architecture

The ITU, in its Recommendation M.3010, defines the principles of a TMN, and the following basic aspects of a TMN architecture, which can be considered separately when planning and designing a TMN:

- functional architecture,
- functional hierarchy,
- physical architecture, and
- information architecture.

The following paragraphs provide more information of these concepts.

### 4.1.2.1 Functional architecture

The *TMN functional architecture* describes the appropriate distribution of functionality within the TMN to allow the creation of building blocks from which a TMN of any degree of complexity can be implemented. The model defines the exchange of management information by means of functional blocks and a set of reference points, as illustrated in Figure 4.1.

In this architecture an operations system function (OSF), accessed by a workstation function (WSF), embodies the principal functions of management and communicates with the network element functions (NEFs) being managed. The interface, referred to as a 'reference point' in the functional architecture, between an OSF and NEFs is called a 'q3' reference point. WSFs and OSFs

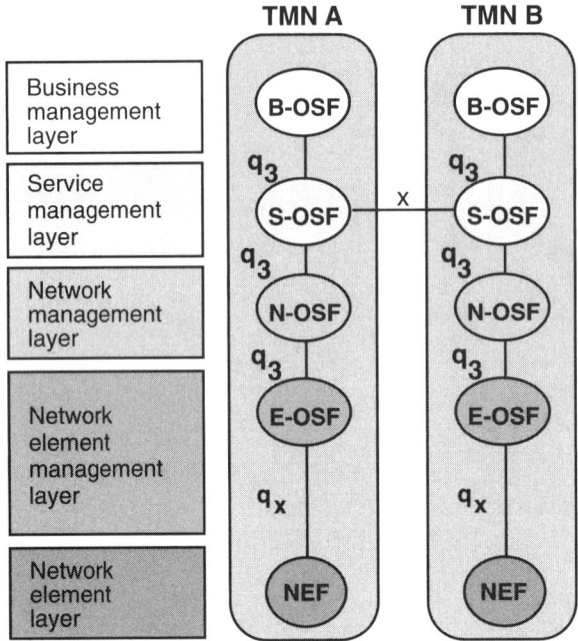

**Figure 4.2** Functional TMN hierarchy.

interact via an 'f' reference point. Interactions between OSFs belonging to different TMN domains are realized via an 'x' reference point. (Note that an 'x' reference point requires increased security beyond the level which is required by a 'q3' reference point.) Sometimes specific functionality is required to permit this communication, and this is located in a *mediation function* (MF). A *Q-interface adapter function* (QAF) connects to TMN those NEFs that do not support standard TMN interfaces.

### 4.1.2.2 Functional TMN hierarchy

Focusing on the management functionality, TMN defines within the functional architecture a functional TMN hierarchy (also known as logical layered architecture in TMN terminology) in which for operational purposes the management is broken down into four layers. Each layer restricts management activities within the boundary of the layer to a clearly defined rank, namely business, service, network, and element management. The element management layer manages each network element on an individual basis, whereas the network management layer has the responsibility for the management of all network elements, both individually and as a set. The network management

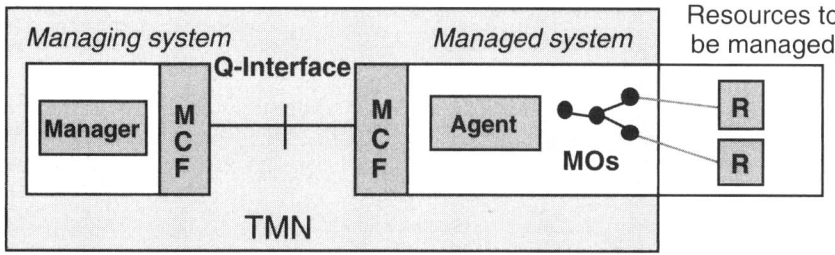

**Figure 4.3** TMN manager–agent model.

layer provides the functionality to manage a network by coordinating activity across the network and supports the 'network demands' made by the service management layer, which is concerned with and responsible for contractual aspects of services that are being provided to customers. The business management layer has the responsibility for the total enterprise and is the layer at which agreements between operators are made.

Each layer is represented by a corresponding OSF, i.e. there is an OSF for business management (B-OSF), an OSF for service management (S-OSF), an OSF for network management (N-OSF), and an OSF for network element management (E-OSF) as indicated in Figure 4.2. Within one TMN all vertical and horizontal interactions take place at generic 'q3' reference points. However, the interactions between the OSFs of different TMNs, i.e. operated by different service providers, take place at 'x' reference points usually at the service management layer.

### 4.1.2.3 Physical architecture

The *TMN physical architecture* describes interfaces and physical components that make up the TMN, in which the management functional blocks are translated into 'physical TMN building blocks'. For example, an OSF becomes an operations system, an NEF becomes a network element, etc. Correspondingly, the reference points are translated into 'interfaces', where 'q3' reference points are realized as 'Q3-type' interfaces and 'x' reference points are realized as 'X-type' interfaces. These interfaces are based on OSI's CMIS/CMIP.

### 4.1.2.4 Information architecture

The *TMN information architecture* describes the nature of information that needs to be exchanged between the functional building blocks. The basic management processes defined in this context are the 'manager' process, responsible for starting management operations, and the 'agent' process, which processes the operations initiated by the managing process on the *managed objects* (MOs). Additionally the agent forwards event reports from the local MOs to the manager, as illustrated in Figure 4.3. MOs are abstract representations of the real resources to be managed, e.g. they model physical components of a network element.

## 4.1.3 TMN applications

Focusing on the TMN applications, ITU Recommendation M.3020 describes a TMN interface specification methodology, that distinguishes between *management services* (MSs), *management service components* (MSCs) and *management functions* as indicated in Figure 4.4. A TMN management service is defined as an area of management activity that provides for the support of an aspect of operations, administration or maintenance (OA&M) of the network being managed. A management service component is the constituent part of a TMN management service stating the requirements for actions to be performed on the managed network (e.g. change customer service details). A management function is the smallest part of a TMN management service as perceived by the user of the service (e.g. report the service availability of network elements).

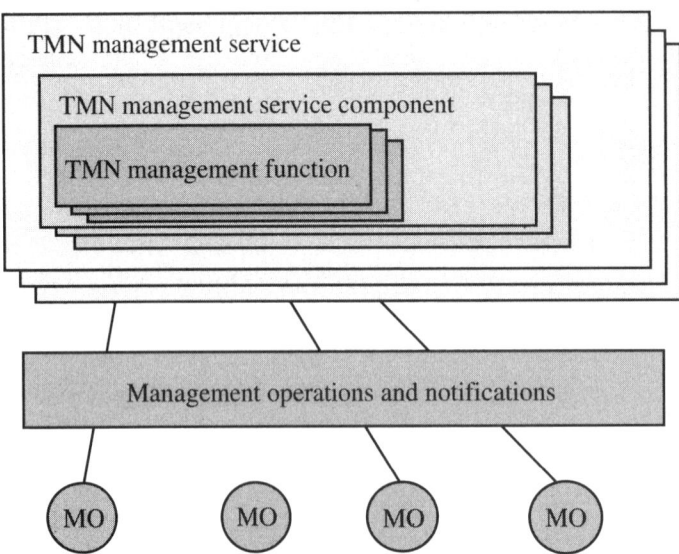

**Figure 4.4** TMN management service decomposition.

The management service components and management functions are related to such aspects as performance, fault (maintenance), configuration, accounting, and security management. Each function achieves its objective using a sequence of actions (management operations) on a defined set of managed objects.

TMN management services are defined in ITU M.3200. Management functions are given in ITU Recommendation M.3400.

Management services (i.e. management functions) residing in an operations system will be achieved by accessing management information, i.e. managed objects, residing in other operations systems or network elements, across TMN interfaces.

## 4.1.4 IN/TMN integration

Both INs and TMNs represent fundamental frameworks for future telecommunications environments. They are closely related, as they cover complementary aspects, i.e. the IN provides for uniform service creation and provision, whereas the TMN enables the uniform management of services and networks (Magedanz, 1993). Nevertheless, the two concepts are not harmonized with respect to functionality, architecture, and methodologies. Consequently, harmonization and integration of the concepts is urgently required for the target telecommunication environment and thus is the subject of several international research and standardization activities. Generally, two evolutionary steps can be identified for that integration:

- medium-term TMN-based management of IN; and
- long term integration of IN and TMN into a common platform.

In the next section we address both of these issues in more detail.

### 4.1.4.1 Medium-term TMN-based management of INs

Obviously the definition of IN management concepts is of pivotal importance for the success of the IN, as the rapid introduction of new services enabled by the IN architecture concept requires the corresponding provision of IN service and network management capabilities. Thus, the application of TMN concepts for the management of IN services and networks in the medium-term time frame is a straightforward approach, as TMN is going to be deployed for nearly all telecommunication systems (IEEE, 1995). In other words, a common management philosophy is being adopted for the management of the IN and other networks (e.g. ISDN, GSM), services and equipment, allowing economies through the use of common techniques.

Research in the area of IN management started in the early 1990s in several international research projects, of which RACE and EURESCOM projects in particular have achieved substantial results (Magedanz *et al.*, 1993; Beires *et al.*, 1994). In parallel, ETSI NA4/NA6 established a joint group of IN and TMN experts in 1989 to study the relationship between INs and TMNs in more detail. The resulting report (ETSI DTRNA-43308) was finalized in 1992 and is the basis for all subsequent IN management activities within the ETSI and also the ITU.

As illustrated in this book, IN standardization initially concentrated on service switching and service control. A deeper study of IN management aspects within the standards bodies has started with CS-2 (Chapter 3). Thus, both the ESTI and the ITU have established joined IN/TMN expert groups in order to define IN management solutions harmonized with both IN and TMN standards (Kockelmans & de Jong, 1995). In the context of TMN standardization, an appendix has been added to ITU Recommendation M.3010, addressing the relationship between INs and TMNs.

As can be recognized from the short description of the TMN philosophy given in the previous subsections, the application of TMN concepts for developing a management solution for a particular telecommunications system is a very complex issue. This is particularly true when studying, in addition to the intra-domain IN management aspects (i.e. IN services provided only within one IN platform), the inter-domain IN management aspects (i.e. IN services provided on interconnected IN platforms).

The starting point for a TMN-based IN management solution is a comprehensive management requirements analysis, covering for each role present in the IN environment (i.e. service subscriber, service provider, network operator) all management activities within the entire lifetime of an IN service, also referred to as 'service life cycle' as defined in ETSI DTR NA-60109. A service life cycle basically comprises a service implementation life cycle (i.e. service preparation, creation, acceptance, deployment, control, and removal), a service subscription life cycle (i.e. service subscription, subscriber control, and removal), and an invocation life cycle. ETSI NA6 has collected a first set of IN management requirements (ETSI DTR NA-60801). The EURESCOM project P226, 'TMN Management of IN-based Services', has contributed substantially to these documents.

Based on the identified IN management requirements, the TMN management solution can be developed, which basically comprises three major constituents (Magedanz, 1994):

1. a set of appropriate IN management services provided to each of the identified IN management roles, i.e. there are specific management services for IN service subscribers, another set defined for IN service providers, etc.;

2. a corresponding IN management information model, providing the basis for the implementation of the above management services; and

3. an IN management architecture, defining the interfaces between the TMN operations systems and the IN network elements and also between operations systems (e.g. between service management operations system and network management operations system).

Taking this into account, it is clear that the IN management functional model developed within CS-2 is only one part of the entire TMN management solution. Hence, in addition to this IN management functional model, corresponding management services and the related management information, i.e. managed objects, have to be defined to achieve a comprehensive TMN-based IN management solution. In this context the TMN standardization is identifying a set of corresponding TMN-based IN management services, which will be defined in the ITU Draft Recommendation M.32IN.

## 4.1.4.2 Long-term integration of IN and TMN into a common platform

The integration of IN and TMN within a common platform allowing the integrated creation, provision, and management of future telecommunication services, comprising both telecommunications and management capabilities, is envisioned for the long term. In this context the ETSI has already performed some studies (ETSI DTRNA-43308) on the integration of IN and TMN concepts in an attempt to develop an enhanced INCM, known as the 'IN management conceptual model'. However it can be assumed that, in the course of defining an integrated IN/TMN platform, it will be necessary to incorporate emerging open distributed processing standards (ITU Recommendation X.900), i.e. the future of telecommunications will be object oriented, with the INCM probably being replaced by the open distributed processing (ODP) viewpoint approach.

In this context new open service architectures, such as Bellcore's information networking architecture (INA) or the international telecommunications information networking architecture (TINA) are gaining momentum. These will be addressed in more detail in Chapter 5. Nevertheless, it has to be stressed that within these new architectures there is no concept of either IN services or TMN services, as these architectures provide a common set of capabilities for implementing both telecommunications and management services. The basic advantage of this idea is that there is no longer any need to separate service control (usually achieved by IN) and service management (usually achieved by TMN). However, at the time of writing there exists a migration path neither from IN to TINA nor from TMN to TINA, although the EURESCOM project P508, 'Evolution, Migration Paths and Interworking with TINA', has started in 1995–96 to study this issue.

Finally, another more controversial thesis should be mentioned, the idea of which is to achieve IN service capabilities by means of TMN concepts (Magedanz, 1995). The rationale for this thesis is as follows. Comparing the increasing scope of emerging TMN (service) management services with the capabilities offered by IN services, there is clearly an overlap of functionality as IN service features focus primarily on the control and management of bearer transmission services (e.g. telephony). The reason for this functional overlap between IN and TMN stems from the fact that most IN service features were designed many years ago, when standardized (service) management concepts were not available but in the face of market needs for enhanced 'bearer' service capabilities and emerging customer control requirements. Consequently the IN could be regarded as a short-term implementation of a *service management network*.

A validation of this thesis has been performed with the implementation of a TMN-based personal communication support system (PCSS), which provides IN-like service capabilities by means of TMN principles to an open set of communication services (Eckardt *et al.*, 1995).

## 4.2 IN and mobile/personal communications

In this section we will address the area of mobile/personal communications and try to illustrate the relationships between IN and mobile communications systems. Thus, we first provide an introduction to the area of mobile and personal communications, then look at the existing mobile communication systems and finally address the relationship between these systems and the IN.

### 4.2.1 Mobile/personal communications

Owing to society's increasing demand for 'universal connectivity' and technological progress, mobile and personal communications are becoming fundamental attributes of future telecommunication systems. Taking into account the vision of future telecommunications, which is 'information any time, at any place, in any form', the trends toward mobile and personal communications can be viewed in terms of three areas.

### 4.2.1.1 Mobility in fixed and wireless networks

Three types of mobility have to be distinguished.

1. Terminal mobility allows users to communicate or obtain access to information services while moving. A terminal is identified by an unique terminal identifier independent of the network point of attachment. The user binds his or her identity to the terminal and hence becomes continuously reachable. This requires that the network must store and maintain location information for the terminal, i.e. the network keeps track of the terminal location.

2. Personal mobility allows users to make and receive calls independent of both the network point of attachment and specific user equipment. This means that a user can use any network access point and any terminal while being identified through a globally unique personal number and charged through the user's personal account. Thus, user-specific location information and services information has to be stored and maintained in a 'service user profile'. Functions for user registration and authentication have to be provided by the service. This type of mobility is a layer above terminal mobility and independent of any radio link.

3. Session mobility allows a service session in which an end user is currently involved to 'follow' that user independent of the location of the user and/or the terminal the user may have access to or of the access arrangement to the network. This requires functions for the service provider to maintain 'session files' containing information about the state and parameters of a session, which are accessible from various locations at the request of the end user.

### 4.2.1.2 Personalization of communication services access and delivery

Personalization describes customers' ability to define their own working environment and service working conditions stored in a 'personal service profile'. This profile defines all services to which the user has access, the way in which service features are used, and all other configurable communication aspects, in accordance with the user's needs and preferences, with respect to parameters, such as time, space, medium, cost, integrity, security, quality, accessibility, and privacy.

### 4.2.1.3 Interoperability of interfaces and services

Interoperability is one step beyond personalization and describes the capacity of a communications system to support effective interworking between different (possibly unrelated) services, supported by and offered on heterogeneous networks, with the long-term aim of achieving fully interworking applications.

## 4.2.2 Overview of mobile communications systems

Today, a jungle of mobile communication systems exists in the telecommunications world. We therefore provide below a short overview of the historical evolution of mobile communication systems. In general, four types of mobile/personal communications systems can be distinguished:

1.  mobile communications systems supporting terminal mobility;
2.  personal communications systems supporting personal mobility;
3.  integrated third-generation mobile communications systems, envisioned to support both terminal and personal mobility, while also integrating mobile satellite systems;
4.  future intelligent communication systems, which in addition to mobility, particularly support personalization of the user's communication environment and communication services interworking.

### 4.2.2.1 Mobile communications systems

Mobile communication systems, supporting terminal mobility, are based on radio mobile networks, which can be categorized into cordless and cellular systems.

First-generation cordless systems, such as *cordless telephone* (CT1), were based on analog technology, replacing the local wire by a radio link. These systems have been followed by digital ones, such as CT2 and *digital European cordless telecommunications* (DECT), which have been designed with two applications in mind, namely wireless access in private (notably office) and in public areas. DECT, operating over the 1880–1900 MHz bandwidth, supports terminal mobility at walking speed and terminal-driven seamless handover. DECT is the new European cordless standard, which will also be used for cordless applications in the public environment, namely for public base stations, known as 'Telepoints' (e.g. Zonephone or Phonepoint in the UK, or Pointel in France). A good overview of cordless systems in general is given in Tuttlebee (1992).

In the context of cell-based systems, the first-generation mobile radio systems were based on analog technology, such as the *American mobile phone system* (AMPS) or the European *Nordic mobile telephone* (NMT). Second-generation radio systems are digital, such as the European *global system for mobile communications (GSM)*, GSM phase 1 being standardized by the ETSIGSM-PN GSM Recommendations. In contrast to previous analog mobile systems, GSM is the first pan-European mobile network, allowing users to roam throughout Europe. GSM was originally developed for vehicular mobile communication, but is nowadays also applied for personal mobile communications as described below.

In contrast to cordless systems, which define only an air interface, a cellular system is much more complex, as it is a complete system, comprising radio interface, infrastructure architecture, network interfaces, and signaling protocols. For example, GSM comprises a radio network for communication with the mobile station (i.e. mobile terminal) and a fixed network portion for maintaining location information of the mobile station in specific real-time databases. The location information stored in these databases, known as 'home location registers' and 'visitor location registers', is accessed for location updates and call set-up via a specific CCS7 signaling protocol, known as 'mobile application part' (section 2.4.3). Hence, many similarities to the IN architecture can be recognized. A good introduction to GSM is given in Mouly & Pautet (1992).

A global cellular system allowing global roaming will be possible with the direct successor of the GSM, the new digital cellular system (DCS1800), also referred to as personal communications network (PCN) in the US. DCS1800, operating at 1800 MHz , defines a derivate of the GSM standard that is designed for higher capacities in urban areas. It allows for smaller cell sizes, higher frequency reuse, and smaller devices and is going to be introduced in the 95–96 time frame on a global basis.

### 4.2.2.2 Personal communications systems

*Universal personal telecommunications* (UPT) is a promising service concept that has been selected as one of the challenging applications of INs and is currently standardized in ITU Draft Recommendation F.851 and ETSI ETR NA-70201. This means that, in contrast to the existing wireless mobile communication systems, UPT implementation will be based on the standardized IN architecture.

UPT enables access to telecommunication services by allowing personal mobility. UPT standardization and introduction should take place in phases: UPT phase 1 is only for support of telephony services over PSTN and ISDN in the 95–96 time frame, whereas UPT phase 2 should allow for broadband services in the long term. In general, UPT should enable users to participate in a user-defined set of subscribed services and to initiate and receive calls on the basis of a personal network-transparent UPT number across multiple networks (fixed or mobile) irrespective of geographic location, limited only by terminal and network capabilities. This means that users have to register explicitly for incoming and outgoing calls in order to provide the system with the necessary location and authorization information. Note that UPT phase 2 will enable users to register with a mobile terminal (i.e. a GSM terminal). A good introduction to UPT is given in Dang *et al.* (1992).

### 4.2.2.3 Future integrated mobile communications systems

In the long term, integration of all mobile radio applications (cordless, cellular, and paging systems), including mobile satellite systems, into one universal system is necessary to support worldwide roaming. The reason for integrating satellite systems into the target system is that satellite systems could complement terrestrial systems in low-density areas. Thus, the ETSI and the ITU are currently working on the definition of third-generation mobile systems, which are expected to start to provide services around the year 2000.

In Europe, the ETSI's Special Mobile Group (SMG 5) is attempting to define a *universal mobile telecommunications system* (UMTS) (ETSI DTR SMG-50301), whereas parallel standardization activities on a worldwide level are being undertaken within ITU-R TG8/1 under the banner of *future public land mobile telecommunication system* (FPLMTS). [Note that FPLMTS has recently been renamed international mobile telecommunications 2000, IMT-2000 (CCIR Recommendation 816; CCIR Recommendation 817).] UMTS and FPLMTS have similar goals and both systems should provide global roaming and inter-system handover capabilities. They should provide access to a wide range of telecommunication services supported by fixed networks (e.g. PSTN, ISDN) and to other services that are specific to mobile users. The systems are therefore expected to be identical or largely compatible.

In contrast to current mobile communication systems, UMTS is targeted to be implemented using a common infrastructure with fixed networks. Full integration of UMTS with B-ISDN, in which UMTS could be considered as the mobile access to B-ISDN, is envisioned in order to support a broad range of teleservices and multimedia services (voice, video, and data). UMTS aims for a maximum user data rate of 2 Mbit/s. To facilitate the fast deployment of UMTS, many functions required for support of mobile systems are expected to be provided by IN CS-2. A good introduction to UMTS is given in Rapeli (1995).

#### 4.2.2.4 Future intelligent communications systems

For completeness it has to be said that, in addition to the mobile/personal communications systems discussed above, new advanced communication systems are emerging, driven by the general convergence of computing and telecommunications and particularly by the convergence of mobile computing and personal telecommunications. In contrast to the above systems, which primarily support mobility, the basic attributes of these new intelligent communication systems is their support of service personalization and service interworking. In this context service personalization allows the definition of advanced screening and call-forwarding scenarios, including the capability of individual announcements for specific callers. Service interworking comprises the definition of advanced service interworking scenarios, such as forwarding all incoming telephone calls to a fax or an email box with or without additional paging of the called party. As stated at the beginning of this chapter these aspects are of fundamental importance for the future vision of telecommunications.

These emerging communication systems offer end users the ability to communicate and to organize their communications environment in accordance with their personal needs, characterized by service, time, space, cost, quality, accessibility, security, and privacy. Hence, these systems can be regarded as an extension of the UPT idea and the IN concept. An extensive description of these approaches is beyond this the scope of book. However, interested readers are referred to Griffith & Velthuijsen (1994), Niebert & Geulen (1994), Eckardt *et al.* (1995), Iida *et al.* (1995), and Rizzo & Utting (1995).

## 4.2.3 Impact of mobility on IN

In the course of this book we have learned that the IN provides powerful call-handling capabilities in a network- and service-independent manner. In mobile/personal communications, the functionality of the IN is responsible for many basic functions, such as location information registration and retrieval, call routing, authentication, charging, and handover. Moreover, the IN is

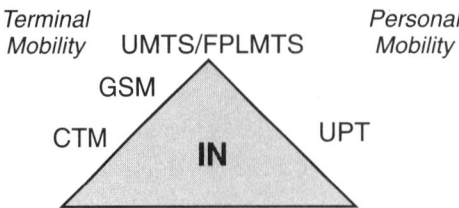

**Figure 4.5** IN providing the basis for mobile/personal communication systems.

expected to add to mobile communication services new advanced calling features, such as personal service profiles and call screening (Martin, 1994).

Current mobile communications systems incorporate functions mentioned above into their own systems, based on 'pseudo' IN architectures. The reason for this is that all of these systems rely on the maintenance of specific subscriber information, such as terminal or user location information, and additional supplementary service information. Such information is held in specific real-time databases within the fixed network part of a mobile communications system and is accessed via specific signaling protocols for location updates and call handling.

This leads to the possibility in future of having different kinds of INs offering similar capabilities within different service environments; this is not desirable from the point of view of network interconnection and seamless service interworking. To avoid this, there is global interest in merging the mobile communication systems with the IN, as the IN provides the required intelligence for such systems within the fixed network in a uniform manner. Since third-generation mobile telecommunication systems, such as UMTS/ FPLMTS, are based on the IN architecture (in fact on an enhanced IN CS-2 architecture!) (ETSI DTR NA-61301), existing mobile systems will have to evolve toward an IN architecture in the long term in the course of their evolution. ETSI DTR SMG-50104 has recently started to study the introduction of UMTS with a look at the evolution of existing mobile telecommunication systems.

As outlined in the previous sections, personal mobility, i.e. UPT, is already based on the IN. Support of terminal mobility is now the focus of IN enhancements (compare with IN CS-2 described in section 3.2). Here we have to differentiate between cordless telephony systems, such as CT2 or DECT, which could be seen as specific 'access networks' to the fixed network, and cellular systems, such as GSM, providing separate (access and core) networks.

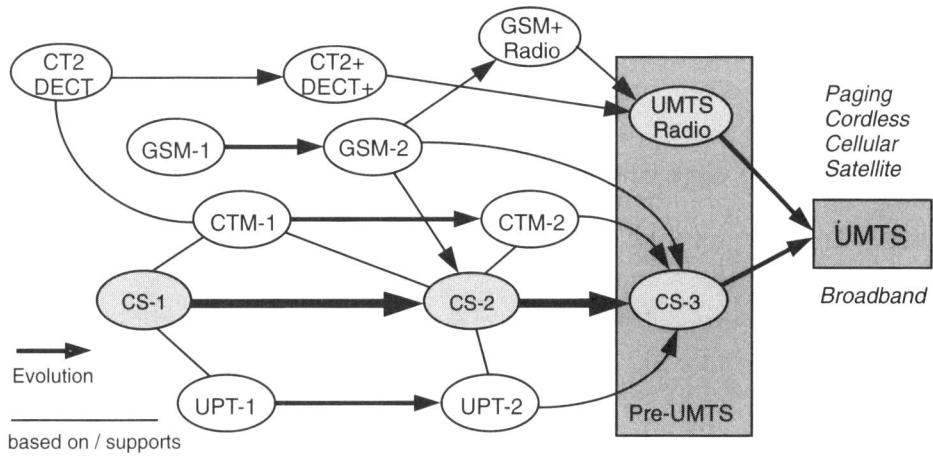

**Figure 4.6** Evolution of mobile communication systems toward UMTS.

Looking at IN support for cordless telephony, the ETSI is currently developing standards for an enhanced IN CS-1 architecture, enabling an European *cordless terminal mobility* (CTM) service, which allows cordless terminals to be connected to public IN-structured networks. The architecture for CTM is defined in ETSI DTR NA-61302, whereas the protocol aspects and the procedures are defined in ETSI Draft DE NA-10039. CTM is defined in phases. CTM phase 1 is aiming at providing intranetwork CTM support for IN-structured public networks and support of CTM internetworking between an IN-structured public network and private telecommunication networks. This means that CTM customers are able to roam between different CT2 or DECT zones by registering in these zones. The CTM service will be independent of the air interface between the cordless terminal and the fixed termination. Thus, both CT2 and DECT air interfaces can be supported. CTM phase 2 will provide enhanced call capabilities, with roaming between CTM and GSM also being envisioned, which could be seen as a basic evolution step toward a pre-UMTS solution.

In the context of IN/GSM integration, ETSI SMG 3 is currently developing a new concept called *customized applications for mobile network enhanced logic* (CAMEL), which should support operator-specific services to mobile users when roaming in visited (mobile) networks. The on-going work concentrates on the definition of the basic CAMEL service features (ETSI SMG3 TCOD 95C101). Furthermore, many research activities are investigating the integration of IN and GSM within future GSM phases, and it can be assumed that the IN architecture will replace the GSM proprietary architecture in the long term. This means that, in principle, CTM and GSM will be based on the same architectural basis, with only the radio interfaces and the radio coverage being different.

In summary it can be stated that the IN concept is the architectural basis for most of the mobile/personal communication systems discussed above, in particular it is the necessary intelligence platform in the fixed network part of mobile communication systems. In fact, it seems to be a general consensus that the success of mobile communications will depend strongly on the support of IN (Figure 4.6). Hence, the development of the second set of IN standards, i.e. capability set 2, is strongly influenced by emerging mobile telecommunication services. This results in the definition of new IN mobility service features (ETSI Draft DTR NA-60902), such as user registration, terminal mobility, and location determination, and the corresponding enhancements of the IN architecture and signaling protocols (ETSI DTRNA-60401) to cope with location management (section 3.2.2). However, it seems likely that comprehensive support of mobility applications could be achieved with CS-3.

Within this book we will not further elaborate on this area, as there are several excellent articles addressing the relationship between IN and mobile/personal communication systems. Thus, interested readers are referred to Hecker (1992) and van Nielen (1992) for more information.

## 4.3 IN and broadband ISDN

The IN is an architectural concept for *all* telecommunication networks. Therefore, in addition to PSTN and ISDN, IN concepts should be particularly applied to the target telecommunications network, known as broadband ISDN (B-ISDN) and the related asynchronous transfer mode (ATM) (Händel, 1994). B-ISDN standardization began in the 1980s and is still in progress. Like IN, B-ISDN is also being standardized in stages, referred to as releases, with three releases currently envisioned. The first standards related to release 1 have matured in the past three years and are available for implementation. The first ATM-based trial networks as well as commercial ATM networks are about to be realized worldwide in different domains, ranging from public backbone networks through corporate networks to customer premises networks.

IN and B-ISDN enhance one another because they provide complementary capabilities with respect to network services, i.e. facilitation of service introduction and flexible service control on the one hand and powerful user network access and call and bearer control facilities on the other hand. Consequently, an integration of both frameworks is of prime importance for both IN and B-ISDN developments. Thus, the integration of IN and B-ISDN is currently the subject of several research projects within ACTS (formerly known as RACE) and EURESCOM. The ETSI has also set up a joint working group for investigating IN/B-ISDN integration (ETSI DTRNA-60108; ETSI DTR NA-60110).

The basic aspects to be studied in the context of IN/B-ISDN integration are described below.

First, new services to be supported by IN in a B-ISDN context, such as broadband video conference services, multimedia services, and video on demand services, have to be defined. In this context the supplementary services already defined within B-ISDN have to be aligned with the existing IN service features.

Closely related to these service definitions is the aspect of call and connection control. Multimedia services, for example, by definition involve multiple media. Each medium may require its own connection, as the different media might be stored in different multimedia archives, resulting in the need for multiple connections in the context of a single call. Consequently, a separation of call and connection control is required. Another example is multiparty conferencing services. These require a distinction to be made in order to isolate functions devoted to call set-up (unique for the duration of a call) from functions associated with connection control (which could vary during a call). For example, during a video conference (representing a call) call parties may join or leave the conference dynamically which means that connections have to be added or removed.

B-ISDN signaling relies on a new concept of separating *call control* and (bearer) *connection control* to enable multiple connections to be set up individually and routed or released during a call. This approach is different from existing signaling systems in PSTN or ISDN, in which connection control and call control are integrated in a so-called 'monolithic' signaling system, allowing only point-to-point connections.

IN CS-1 standards contain no separation between call and connection control. However, in IN CS-2/CS-3 this separation of call/connection control is anticipated. This will result in the definition of a corresponding IN basic call/connection state model, as well as an integrated IN/B-ISDN functional model, which defines corresponding new functional entities in the IN DFP. These are a *bearer control agent function* (BCAF) and a *bearer control function* (BCF) in addition to the already defined call control agent functions (CCAF) and call control function (CCF).

Additionally, some existing IN functional entities, particularly the service switching function and the specialized resource function (Leconte, 1995), have

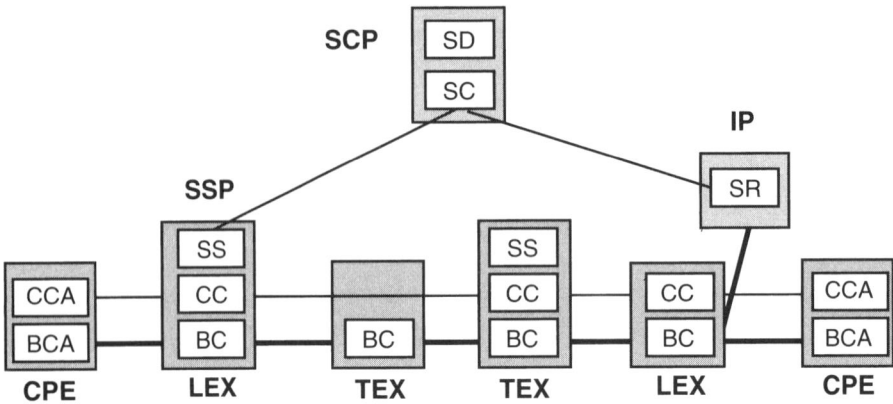

**Figure 4.7** Separation of call control and connection control within the IN functional model.

to be enhanced to support B-ISDN services. Figure 4.7 depicts an enhanced functional IN architecture that is currently under discussion. Within this figure we adopt the B-ISDN notion of customer premises equipment (CPE) for the user terminals and the separation of local exchanges (LEXs) and transit exchanges (TEXs). However, for simplification we do not distinguish between originating and terminating exchanges.

Furthermore, the separation of call and bearer (connection) control will also have great effects on the IN signaling protocol (i.e. INAP), which has to be aligned with the emerging B-ISDN signaling protocol (broadband ISDN user part, BISUP).

It should be mentioned that third-generation mobile systems as defined, such as UMTS/FPLMTS (see above), which will adopt IN techniques for the support of mobility procedures (e.g. location management and handover) and antici-pate B-ISDN for the fixed network access part, are considered as an important application area for an integrated IN/B-ISDN framework.

In summary, it has to be recognized that the integration of IN and B-ISDN is an issue of ongoing research. For more information on this subject interested readers are referred to Bretecher & Vilian (1995), Lauer *et al.* (1995) and Minzer *et al.* (1995). Furthermore, TINA's connection management architecture (TINA, 1994a) is also considered to represent one possible approach for the integra-tion of IN and B-ISDN (Chapter 5).

# 5 Beyond IN

As mentioned above, there are other initiatives besides the international standardization bodies that investigate the long-term evolution of the IN, taking into account the emerging impacts of international *telecommunications management network* (TMN) (ITU Recommendation M.3010) and *open distributed processing* (ODP) (ITU Recommendation X.900) standards (Figure 5.1). The most important initiatives in this context, gaining worldwide attention, are Bellcore's *information networking architecture* (INA) and the international *telecommunications information networking architecture* (TINA).

The targets of both architectures are very similar in the respect that they are designed to support 'open' telecommunications, i.e. both architectures address the needs of telecommunications services, ranging from traditional voice-based services to interactive multimedia, multiparty services, information services, as well as management services. All these services are considered to be software-based applications that operate on a distributed computing platform. This platform hides from applications the underlying technologies and distribution concerns, thus supporting the construction of portable and interoperable code.

It is important to note that neither architecture aims to provide backward compatibility with any existing technology, i.e. INs or TMNs, as the starting point for both architectures was to make use of recent advances in the fields of distributed computing. Thus, these architectures incorporate object-oriented design principles for service modeling and implementation in order to improve

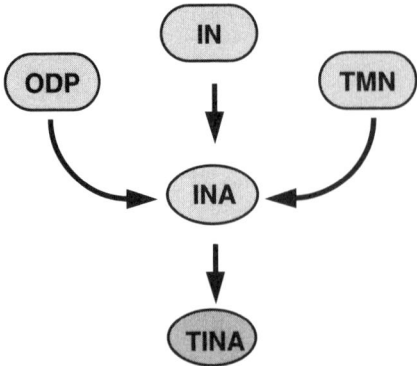

**Figure 5.1**   IN long-term evolution.

interoperability, reuse of software and specifications, and flexible placement of software on computing platforms/nodes. Note that this is in contrast to the function-oriented modeling of IN services based on the INCM. This means that, within these new architectures, we neither recognize any service features and SIBs nor find any IN network elements.

In contrast, we will recognize within both architectures a 'distributed processing environment', collectively provided by the network nodes. This distributed processing platform enables interactions between distributed objects, which collectively achieve a specific telecommunications or management service. This means that a service is achieved by a set of interacting objects that are distributed across different network elements, with the distributed processing environment providing distribution transparency. Following the object-oriented approach new services can easily be achieved by enhancing already existing objects or defining additional objects.

In the next section we consider both architectures in more detail, with the emphasis on TINA. We presume that readers have a minimum knowledge of object-oriented programming in order to understand the information given in this chapter. Readers without this knowledge should refer to Rumbaugh *et al.* (1991) for the principles of object-oriented programming.

## 5.1 Information networking architecture

Bellcore in 1990 initiated a work program on the long-term IN for the time beyond AIN, referred to as *information networking architecture (INA)*. An information network is defined as a telecommunication system that provides

Table 5.1 INA specifications

| Document no. | Document title |
| --- | --- |
| SR-NWT-002280 | *An Introduction to the INA Field Experiment Initiative*, Issue 1 |
| SR-NWT-002281 | *INA Cycle 1 Documentation Road Map*, Issue 2 |
| SR-NWT-002282 | *INA Cycle 1 Framework Architecture*, Issue 2 |
| SR-TSV-002283 | *INA Cycle 1 Contract Specification*, Issue 2 |
| SR-NWT-002284 | *INA Cycle 1 Distributed Processing Environment Specification*, Issue 2 |
| SR-TSV-002285 | *INA Cycle 1 Trading and Naming Specification*, Issue 2 |
| SR-NWT-002286 | *INA Cycle 1 Network Management Functional Architecture*, Issue 2 |
| SR-NWT-002287 | *INA Cycle 1 Management Information Model*, Issue 2 |
| SR-NWT-002288 | *INA Cycle 1 Service Management Architecture*, Issue 2 |
| SR-NWT-002289 | *INA Cycle 1 Protocol Specification*, Issue 1 |
| SR-TSV-002290 | *INA Cycle 1 Security Specification*, Issue 1 |
| SR-TSV-002291 | *INA Cycle 1 Data Management Specification*, Issue 1 |
| SR-TSV-002660 | *INA Cycle 1 Communications Management Architecture*, Issue 2 |
| SR-NWT-002804 | *INA Cycle 1 Contracts and DPE Experiments Report*, Issue 1 |
| SR-NWT-002806 | *INA Cycle 1 Configuration Experiments Report*, Issue 1 |

access and management of information at any time, any place, in any volume, and in any form. INA defines a set of concepts and principles for the development and provision of telecommunications application software. It must be stressed that the functionality of software components in an INA-consistent network include functionality that traditionally belongs to network applications (i.e. IN services) as well as functionality that traditionally belongs to operations applications (i.e. TMN services).

INA therefore focuses on the definition of a framework architecture that is based on modern software concepts, such as distributed computing and object orientation, and integrates current work in the field of INs, TMN, and ODP. The keyword is 'interoperability' of INA application software. In the course of the INA activities Bellcore has developed an INA *distributed processing environment* (DPE) for public telecommunication networks, known as 'INAsoft DPE' (Natarajan, 1995), which should support the first INA applications. The first set of INA specifications, known as *Cycle 1 Specifications For Information Networking Architecture* (INA) (Issue 1) was released in June 1992 (Bellcore, 1992c); an enhanced version (Issue 2) was published in April 1993 (Bellcore, 1993c). The INA documents in Table 5.1 are available. INA specifications can be obtained from Bellcore Customer Relations (see p. 109).

It seems likely that there will be no further INA specifications in the future, because another international research initiative in the field of long-term IN, strongly influenced by INA results, was set up in 1993 under the banner of *telecommunication information networking architecture* (TINA), which is described below. Thus, we do not describe the INA concepts within this book. However, interested readers should refer to Natarajan & Slawsky (1992) and Barr *et al.* (1993) for an overview of INA.

## 5.1.1 INA principles

An important concept in INA is the separation of network functionalities into two so-called segments, namely:

1. a *service segment* providing functionalities dealing with the provision of network services to users; and
2. a *delivery segment*, which includes functionalities for switching and transmission of information dependent on the specific technology used in transmission and switching equipment.

This segmentation principle identifies a major interface between these two segment types, for which a manager–agent model plus the corresponding managed object model from OSI management standards is adopted. Note that interfaces of this kind would typically be used even for call handling-related processing, whereas today they are typically used only for management purposes. This separation principle can be considered as the ultimate evolution of the unbundling between service control logic and data from basic switching and transport functions, first advocated by the IN concept. INA generalizes this principle to the inclusion of management, i.e. for customer network control, to the call control component of a network service.

Another major architectural concept in INA is structuring the service segment and delivery segment in software units called *building blocks*, borrowed from the OSCA (operations system computing architecture), in course of development in Bellcore. Building blocks are built, installed, and maintained as separate units, and interoperate via invocation of operations on their published interfaces according to client–server and, in some cases, distributed transaction-based mechanisms. A key point is here the possibility of installing different building blocks provided by different vendors or directly developed by the operating company. A building block is thus the unit of operability, distribution, security, and interoperability. For such purposes the building blocks are categorized into different types (layers) depending on the functionality they provide based on the OSCA functional separation. These are the data layer, guarding corporate data; the user layer, dealing with human user interfaces; and the processing layer, supporting business processes. Note that the building blocks must not provide both delivery segment and service segment functions. A service segment building block provides only service segment functions, a delivery segment building block provides only delivery segment functions.

The structuring of application software within building blocks should be object based, to promote modularity, encapsulation, and reusability advantages associated to the object-oriented paradigm. A service provided by an object contained in a building block and visible outside of that building block is called a *contract*. Two objects that are located in different building blocks interact only via contracts. It should be noted that this notion of objects encompasses both programmatics, such as those present in OSI management CMIS/CMIP, and non-computational objects, such as a communication path.

An essential ingredient of the INA is the *distributed processing environment* (DPE), on which service segment and delivery segment building blocks are executed. The DPE can be considered to be composed of two major layers: the DPE kernel, present at every node, and a collection of DPE servers for functions such as interface trading, transaction support, and security control.

By analogy with ODP these servers are implemented with system building blocks. For both layers emerging '*de jure*' and '*de facto*' standards are adopted whenever possible.

The INA service architecture prescribes rules for functional structuring of the applications in modules (to be developed as sets of building blocks in different nodes) and their interactions. Currently, this service architecture is based on an original extension of the OSI management framework and its system management functional areas (configuration, fault, performance, accounting, security) to include connection control and a combination of this partitioning of functions in a three-level hierarchical model encompassing service-, network- and network element management, according to the TMN functional hierarchy.

## 5.1.2 INA architecture

Software functionality within an INA network node may be divided into different levels, all of which exist above the hardware comprising the node, as illustrated in Figure 5.2. The lowest level forms the *native computing and communication environment* (NCCE), which consists of the operating system and other related services, such as database management systems, that form the computing environment and transport layer services that provide end-to-end

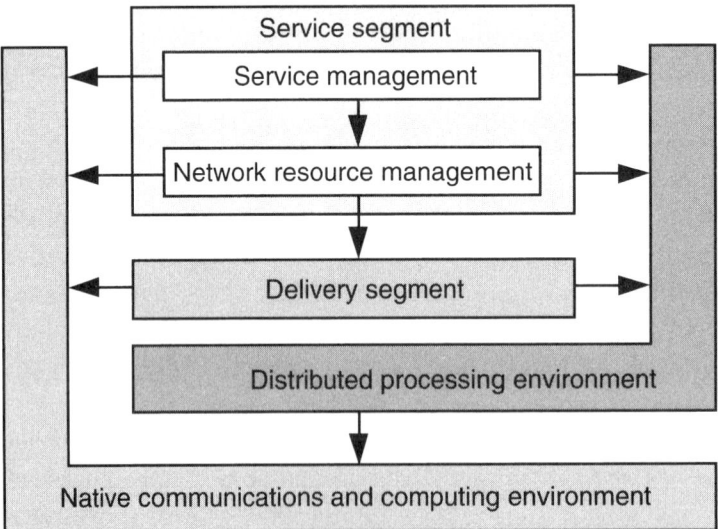

**Figure 5.2**  INA architecture levels.

communication services. Unlike other architectures, the network layer (beneath the transport layer) does not include routing and congestion control functions, which are performed by service segment building blocks at the network resource management functions level.

At the next higher level we have the distributed processing environment (DPE), which provides services that enable distributed processing. It provides the infrastructure for building blocks to interact via contracts and provides distribution transparencies for these interactions. It does not manage network resources! The DPE is composed of the DPE runtime environment, comprising DPE kernel and DPE servers, and application development support tools such as contract adapter generators and contract dictionary servers.

The third level is called the delivery segment functions level, which consists of delivery segment building blocks. These hide the product technologies of switching and transmission equipment, and provide management operations on a set of managed objects. Thus, these building blocks correspond to the agents in the OSI management/TMN context (section 4.1). The operations are invoked via contracts that are specified by identifying the set of managed objects handled by the contract and functional units (representing a grouping of management operations) applicable to that set of managed objects.

The next two higher levels, namely the network resource management functions level and the service management functions level, constitute the service segment.

The network resource management functions level provides services for the management and control of network resources. Functions provided in this level are independent of the services provided to end users and are organized on the basis of the OSI system management functional areas (i.e. configuration, fault, performance, accounting, and security management) plus a new one called 'connection management', which is considered to be important for network resource management.

The service management functions level is composed only of service segment building blocks, which provide end user service-specific functions, including the service logic and management functions specific to end user services. All interactions with human users occur only at this level. Note that the arrows in Figure 5.2 denote which level uses the services of which other levels. The service offered by the NCCE and DPE services offered by the DPE are not specified as contracts, whereas services offered by other levels and DPE servers are specified as contracts.

Basically service and network management in INA can be distinguished by their management focus, the type of objects that each manages, and the complexities they attempt to hide from their users. Network management focuses on the network and entails the management of the entities that provide the transmission and switching capabilities of the network. Network management shields service management from the underlying transmission and switching facilities needed to provide an information or communication service. Service management is focused external on the network and on providing services to users. Thus, network management provides a unified resource view across many network resources (which employ many network technologies) while hiding the complexity of the resources from the service developer. Service management hides the complexities of service configuration, maintenance, monitoring, and evolution from the user.

In summary, one can say that the INA is based on the current advances in broadband communication and distributed computing technologies and specifies an application architecture framework for future information networking applications, such as multimedia information services. The major principle of INA is the separation of service functionality (service segments) and network functions (delivery segments), which is consistent with the manager–agent separation of current network management standards. Additionally, the rigid division between network applications and management applications has been eliminated because in INA both kinds of applications will be executed on a common distributed processing platform. This platform hides the effects and complexities introduced by distribution from the applications.

# 5.2 Telecommunication information networking architecture

The TINA consortium (TINA-C), comprising all major network operators and computer vendors, was founded in 1992. It aims for the definition of an overall information networking architecture, known as *telecommunications information networking architecture* (TINA), to be used for any type of network (such as PSTN, ISDN, B-ISDN) for both telecommunications and operations applications. The intention is that all applications (telecommunications, operations, and management) will be easier to develop and maintain within an environment in which global aspects increasingly have to be taken into account. The work of TINA-C is strongly influenced by Bellcore's INA specifications, but TINA-C has a broader scope of work (Rubin & Natarajan, 1994).

The focus of current work is the development and validation of architecture specifications, which should be based as far as possible on an integration of

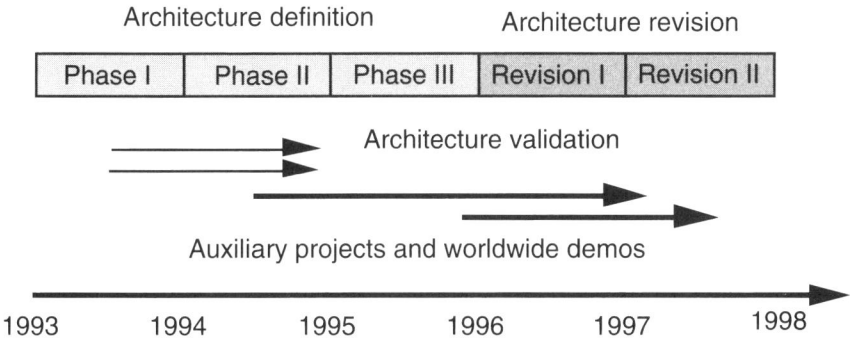

**Figure 5.3(a)** TINA specification schedule.

available concepts, standards, and products. A hardware development is not planned; software development concentrates on the development of prototypes for field trials. This will be accomplished by so-called TINA-C auxiliary projects. Like INA, TINA-C aims for the definition of a framework architecture (Nilsson *et al.*, 1995) and the specification of a TINA-C distributed processing environment (Kelly *et al.*, 1995). The TINA-C architecture should provide the basis for a broad spectrum of services, enabling in particular broadband communication, multimedia, and mobility services.

The target TINA specifications should be available at the end of 1997. On the way toward this target, several interim architecture releases are foreseen (Figure 5.2). The first set of TINA-C specifications was finalized in 1993 (phase I), and a second set became available in spring 1995 (phase II). (Until 1995 TINA specifications were considered to be confidential to the TINA-C member companies. However, the most important specifications mentioned in this chapter are now available for the general public. These specifications can be obtained from the TINA-C web server 'http://www.tinac.com/'.) A third version is anticipated for 1996 (phase III). Furthermore, architecture revisions are scheduled for the end of 1996 and 1997. So-called 'TINA auxiliary projects', which have begun in parallel with the core team work, are considered major contributors to the validation and revision of the TINA specifications (Figure 5.3(a)).

In the following sections we provide more information on the TINA and address first the overall TINA and then concentrate on the TINA service architecture. However, before we look at the TINA, we briefly introduce the principles of the open distributed processing standards that represent the foundation for TINA.

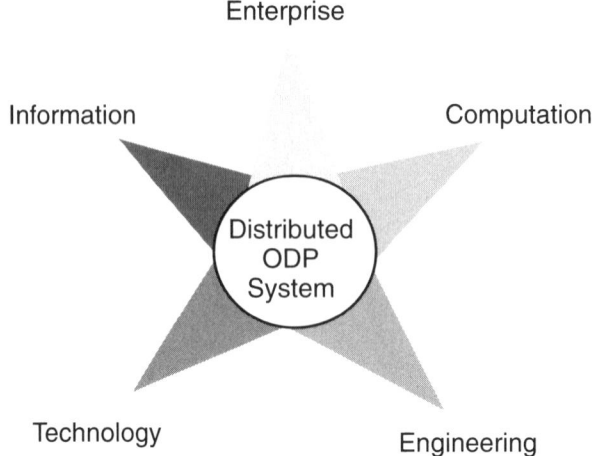

**Figure 5.3(b)**   ODP viewpoints on a distributed system.

## 5.2.1 A short introduction to open distributed processing principles

Open distributed processing (ODP) is a long-term research and standardization initiative in ISO aimed at developing distributed processing systems based upon standards. The core of ODP activities is the definition of a *basic reference model for ODP* (RM-ODP) (ITU Recommendation X.900) that defines an architecture to guide development of specific standards components for ODP. The scope of ODP encompasses all potential applications of distributed processing systems, including (tele)communication services, networks, and their support environments.

Regarding the structure of the ODP system, there are a number of different 'roles' that have interest in any given information processing system. Examples include managers who commission the system, architects who design it, programmers who implement it, and technicians who install it. Each person is interested in the same system, but their relative views of the system are different: they see different issues, they have different requirements, and they use different terms to describe the system.

The ODP reference model (RM-ODP) attempts to recognize these different interests by defining five *viewpoints*, each of which provides a full description of abstractions of a complete system that emphasizes a particular set of concerns (Figure 5.3(b)). In general, the issues visible from each viewpoint are as follows.

- *Enterprise viewpoint:* focuses on business requirements, structure and organization, management policies, human user roles with respect to the open distributed processing system, and the environment with which the system interacts.
- *Information viewpoint:* addresses information sources, sinks, repositories in the open distributed processing system, and information flows between them. The information structure and exchange aspects are modeled.
- *Computation viewpoint:* regards the distributed processing in terms of system components (e.g. application components) and their interactions independent of any specific operating or communication system. It involves the operational and computational characteristics of the distributed processing system, and the processes which exchange the information.
- *Engineering viewpoint:* concerns configurations, performance, distribution, and the infrastructure. It sees a distributed processing system in terms of system services, communication components, and engineering mechanisms that enable the distribution of programs and data.
- *Technology viewpoint:* encompasses hardware and software components, such as I/O devices. It concentrates on technical artefacts (realized components) from which the distributed processing system is built. It must model the hardware and software that constitute the local operating system, I/O devices, storage, etc.

Each viewpoint corresponds to a particular abstraction of an ODP system. Note that there is no link with phases in a design process, sequence, or hierarchy.

Using the ODP viewpoints to structure the analysis of distributed systems problems leads to a clear separation of concerns and consequently a better understanding of the issues involved in their interrelationships. The goal of the RM-ODP is to characterize the problems addressed by each viewpoint and to develop standardized solutions. This will lead to the development of ODP-conformant software components, including both system components and application components. Thus, the RM-ODP provides the architectural framework within which to reason about such components, and their relationships to one another.

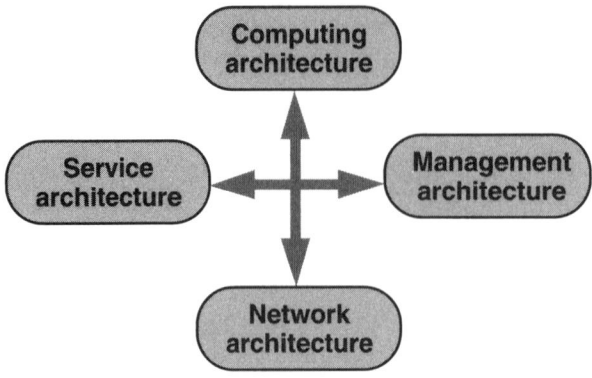

**Figure 5.4** TINA architecture subdivision.

## 5.2.2 TINA overall architecture

Basically the TINA addresses a wide range of issues and provides a complex set of concepts and principles (TINA-C document number TB_MDC.018_1.0_94; TINA-C document number TB_MH.002_2.0_94). In this respect it is much more a framework than a specific architecture. In this context it has to be stressed again that there is no correspondence with the INCM introduced in this book. Rather, TINA is a compilation of state-of-the-art concepts and principles for developing future distributed telecommunications and management services. Among others, ODP principles represent the basic foundation for TINA, in which the TINA specifications concentrate primarily on the information, computation, and engineering viewpoints.

In order to handle the complexity, the overall TINA architecture is structured into four areas (as depicted in Figure 5.4).

1. *Computing architecture* defines a set of concepts and principles for designing and building distributed software and the software support environment, based on object-oriented principles. This architecture is based on the reference model for open distributed processing (RM-ODP) and distinguishes five viewpoints for distributed system specification: enterprise, information, computational, engineering, and technology viewpoints.

2. *Service architecture* defines a set of concepts and principles for the design, specification, implementation, and management of telecommunication services. Three main concepts can be identified: *session*

concepts, which address service activities and temporal relationships; *access* concepts, which address user and terminal associations with networks and services; and *management* concepts, which address service management issues.

3. *Network architecture* defines a set of concepts and principles for the design, specification, implementation, and management of transport networks. Basic aspects of this architecture are the definition of a generic network resource information model, the definition of connection graphs providing a service-oriented view of connectivity, and the connection management. Connection management is the computational model for the establishment, modification, and release of connections in the resource layer.

4. *Management architecture* defines a set of concepts and principles for the design, specification, and implementation of software systems that are used to manage services, resources, software, and underlying technology.

In addition, an overall architecture defines generic concepts and principles for the design, specification, and implementation of any type of software system in a TINA-consistent way. In the next section we concentrate on the computing and service architectures.

## 5.2.3 TINA computing architecture

The TINA computing architecture defines the modeling concepts that should be used to specify object-oriented software in TINA systems and thus provides the foundation for the other three architectures. These concepts are based on the five viewpoints identified in the RM-ODP, namely the enterprise, information, computational, engineering, and technology viewpoints. However, TINA concentrates primarily on the information, computation, and engineering viewpoints.

*Information modeling* concepts are defined in TINA-C document number TB_EAC.001_1.2_94. In this activity the information a service or service component needs is modeled. This information is collected in *information objects*, their mutual relationships, and their constraints and rules that govern their behavior. As notation language, the TINA core team introduced the so-called 'quasi-Guidelines for the Definition of Managed Objects/General Relationship Model' (quasi-GDMO/GRM) notation. Additionally, the *object modeling technique* (OMT) has been adopted for graphical notation.

Operational interface notation

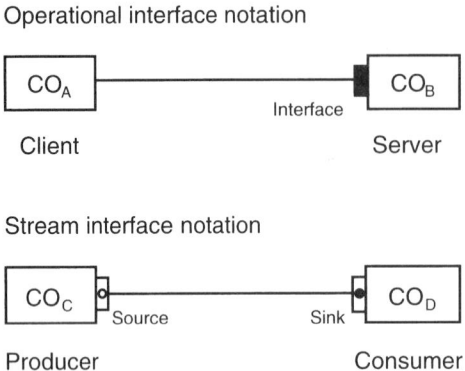

Stream interface notation

**Figure 5.5** Computational object interface types.

*Computational modeling* concepts are defined in TINA-C document number TB_A2.HC.012_1.2_94. Within this activity the application is described in terms of interacting *computational objects*. Computational objects are the units of programming and encapsulation. Objects interact with each other by sending and receiving information to and from interfaces. An object may provide many interfaces of either the same or a different type. There are two forms of interfaces that an object may offer or use: *operational interfaces* and *stream interfaces* (Figure 5.5). An operational interface is one that has defined operations that allow for functions of the offering (server) object to be invoked by other (client) objects. An operation may have arguments and may return results. A stream interface is one without operations. The establishment of a stream between stream interfaces allows for the transfer of structured information, such as voice or video bit streams.

Computational objects are specified with the TINA *object definition language* (ODL) (Kitson *et al.*, 1995), which is an extension of the Object Management Group's (OMG's) interface definition language (IDL) (Kitson, 1995).

*Engineering modeling* concepts are defined in TINA-C document numbers TB_NS.005_2.0_94 and TR_KMK.001_1.1_94. These concepts define the abstract TINA-C machine, i.e. the model of the TINA *distributed processing environment* (DPE). The DPE provides the necessary platform for the computational objects, allowing these objects to interact with each other without having to know on what computer or network node the other object resides. From an application designer's viewpoint the DPE can be considered as one homogeneous infrastructure hiding the complexity and heterogeneity of the underlying network

Interacting service components

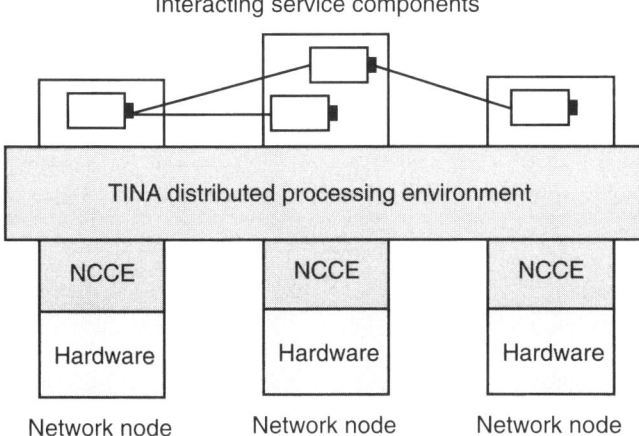

**Figure 5.6** TINA distributed processing environment.

and computing resources (Figure 5.6). However, in reality a DPE consists of 'kernels' implemented on top of different heterogeneous computing environments, so-called *native computing and communications environments* (NCCEs). The DPE kernels are enhanced by additional generic software components, called DPE services, dealing with distribution, security, quality of service, operability, etc. Consequently, the DPE provides the abstraction from the distribution of network nodes and thus the common basis for the distributed realization of TINA services.

At the time of writing there exists the common opinion that the Object Management Group's (OMG's) *common object request broker architecture* (CORBA) provides a suitable basis for a TINA DPE implementation (Kitson, 1995).

## 5.2.4 TINA service architecture

In this section we focus on the most interesting aspects of the service architecture (Berndt *et al.*, 1995; TINA document number TP_MDC.012_2.0_94), namely the session and access concepts. In particular, the access concept provides a generic mechanism for TINA services access. It must be stressed that this description is not complete and highlights only the basic concepts. In general, the service architecture defines fundamental concepts and principles with which TINA service components should comply.

### 5.2.4.1 Session concept

As TINA is expected to support even complex multimedia telecommunication services, the notion of a 'session' was introduced, replacing the traditional concept of a 'call'. A session is the purpose of a service that is achieved by performing a collection of activities during a specific period of time. From the information viewpoint four types of sessions have been defined (Figure 5.7).

1. *Service session* is the service activities and functionality that have network-wide impact. It provides a user or a group of users with an environment to support the execution of a service.

2. *Access session* supports the user in accessing, requesting, and retrieving services or already active services (sessions). Its initial configuration is described at subscription time. The access sessions capabilities can be seen as the starting point for allowing interactions between the user and the available set of services. Furthermore, the access session creates and manages all service sessions in progress.

3. *User (service) session* is the service activities having an impact on a single user. It comprises the activities performed and the resources allocated by one user for one specific service session. Thus, it reflects the settings and constraints imposed by the user or his or her end system (e.g. terminal limitations). Hence, the user session addresses resources to convert information flows and supports specific human interfaces. It is created when a user joins a service session; it is deleted when the user leaves.

4. *Communication session* provides an abstract view of connection-related resources and supports the activities needed to establish the communication between users.

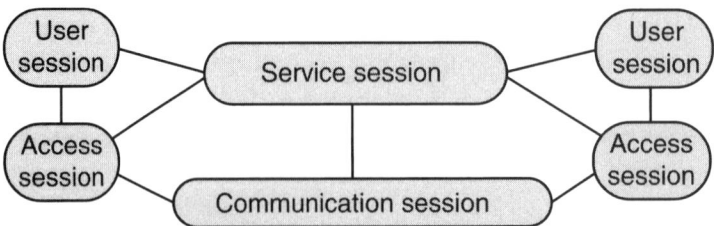

**Figure 5.7** TINA session concept.

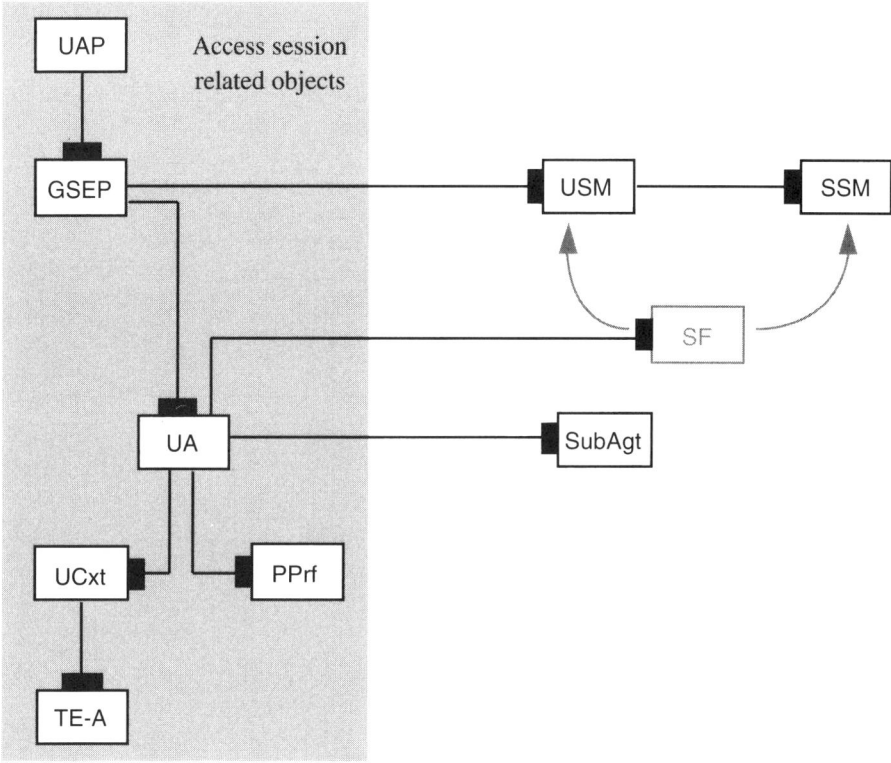

**Figure 5.8**   Access session-related computational objects.

Communication sessions and access sessions are service independent, while service sessions and user sessions are service dependent. The purpose of these session concepts is to separate different concerns and to promote distribution of functionality. The separation of access and service sessions allows both the access methods and the technology for different users to vary, and the location of users accessing the service to change whilst a service is in progression. The separation of user and service sessions allows for the distribution of functions and state, whereby the user session provides a local view, and the service session provides a collected view. This separation supports the suspension and resumption of service involvement. The separation of service and communication sessions supports the division of activities of the service from the set of connections that exist.

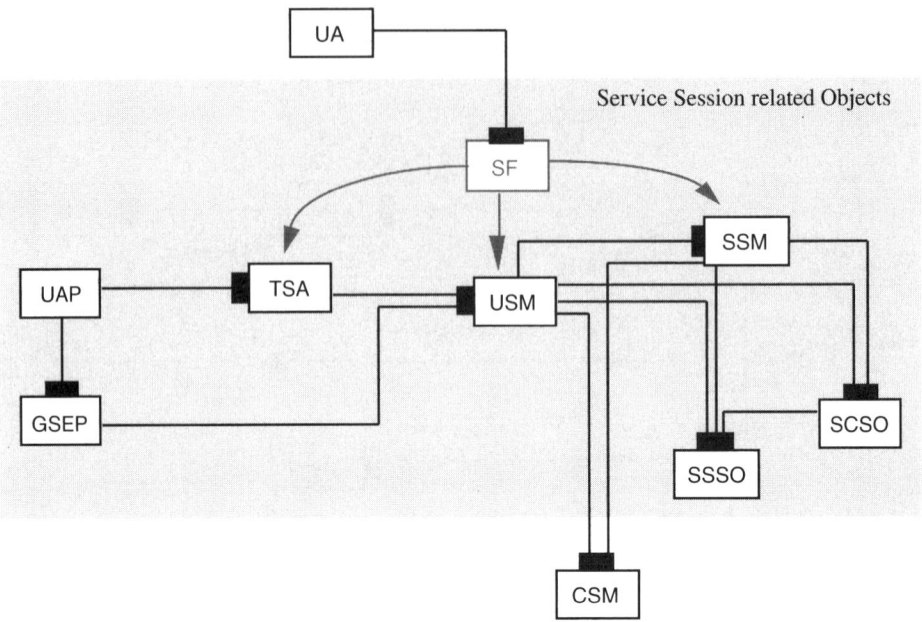

**Figure 5.9** Service session-related computational objects.

### 5.2.4.2 TINA service components

Services are realized within TINA by a set of interacting components, i.e. objects within a distributed processing environment. The TINA service architecture therefore introduces a generic set of components for telecommunications services. As outlined in the computing architecture, TINA separates *information objects* (IOs) from *computational objects* (COs), the latter offering an operational interface. In many cases, information objects are mapped to a corresponding computational object, encapsulating the information in order to be able to define an operational interface to distributed information. This section describes the main computational objects of the service architecture related to the access session (Figure 5.8) and the user/service session (Figure 5.9).

*Service components related to the end user system*

- *UAP*: The user application is defined to model a (variety of) service application in a user system. It acts as an end point of a service session offering the applications user interface. Application stream interfaces can be bound to those in other user systems or service provider servers by communication session managers.

- *GSEP*: The generic session end point is a service-independent computational object and models the minimal set of capabilities as an end point of an access session by interacting with the user agent to perform service session control. A GSEP gathers information about identification and registration of a user and pass it to the user agent in order to access services, conveys requests from the user to the user agent for creating or joining service sessions, receives invitations to existing service sessions from the user agent, and conveys requests from the *user session manager* (USM) to invite other users to existing service sessions. The aggregation of UAP and GSEP is called an end user system.

The UAP as well as the GSEP are involved in both access and the user/ service sessions.

*Service components related to the access session*

- *UA*: The user agent is a service-independent component representing a user in the service provider domain. It acts on behalf of the user and may be seen as a simple intelligent agent-like component. The user agent controls and manages user sessions as well as service sessions, and thus it is the contact point of control for session creation, suspension, resumption, and deletion. Each time a user session or a service session has to be created, the user agent consults a service specific service factory to generate the session performing computational objects (USM or both USM and SSM). Before gaining access to services and configurations, the UA performs authentication and authorization checks, the latter by using an interface component of the subscription management (subscription agent), applying restrictions set by the service subscriber (service customer). Moreover, the user agent allows for personal mobility by maintaining the terminal registration container usage context (UCxt, see below). In a registration scenario, the user agent acts as a receiver of terminal registration messages, which it relays to the user's usage context. In an invitation scenario, it queries the usage context for an appropriate user end system at which the user is registered in order to deliver an invitation by alerting the user at its terminal. The user agent is also responsible for service customization with restrictions and preference given by the users themselves. It uses a personal profile (PPrf, see below) that carries service generic preferences on service execution.

However, in addition, the user agent is the single point of control for maintaining the personal profile, for example modifying users' preferences (customization).

- *PPrf*: The personal profile maintains the user-related constraints and preferences on service access and session execution. These settings determine the environment in which the service will be executed for the user. Basically, the personal profile contains generic constraints, generic service customization data, and service generic usage preferences.

- *UCxt*: The usage context maintains the pool of resources available to the user for the execution of services. It contains registrations at user terminals and terminals at network access points. It is referred to when determining whether there exist available resources satisfying the requirements of a specific service in case of a service creation task or a specific session invitation. The usage context plays an important role in personal mobility support for a user as it keeps track of terminals and access points being used or available for the user. The UCxt interacts with terminal agents (TE-As) to obtain the terminal capabilities with respect to required quality of service (QoS) parameters.

- *TE-A*: The terminal equipment agent is a terminal of a user system within the service provider domain. It maintains the capabilities and state of a terminal from the provider's perspective. It may be created by service subscribers or users.

- *SubAgt*: The subscription agent is a contact point for accessing subscription information for user agents (e.g. a list of available services and a set of constraints posed by a service provider and a subscriber to invoke the service). It interacts with other subscription-related service components, which will no be considered within this book.

*Service components related to the service session*

- *SSM*: The service session manager supports service capabilities that are shared among users in a service session. It keeps track and control of resources (e.g. by having references to other computational objects such as the communication service manager), keeps the session's state and supports suspension and resumption of the service session. Moreover, the SSM supports adding, inviting, and removing of users to the service session and offers negotiation capabilities among users interacting with the USMs. Finally, the SSM supports management capabilities associated with the service session (e.g. accounting). An SSM is created by a service factory (SF) when a service is invoked through an access session. The life cycle of the SSM is determined by the service session controlled by an access session.

- *USM*: The user (service) session manager comprises the service and session control of the user (service) session. It represents and holds the context of a user in a service session. It offers the user the interface of access to service capabilities, it contains local information and local service capabilities for the user, it interacts with user applications in the user terminals, it keeps track and control of the non-shared resources (e.g. by having the reference of communication service manager), it holds the state and supports suspension and resumption of the user session, it supports the negotiation capabilities for the user, and it supports different roles of the user in the service. A USM is created when a user joins a service session and deleted when the user leaves the service session. Thus, its life cycle is bound to the user session as part of the service session.
- *TSA*: The terminal service adapter adapts the user service provider interface (USPI) operations that are offered by the USM into user interface actions (e.g. X Windows System adaptation). It depends on both the specific service and the terminal.
- *SCSO*: A service core support object models service control capabilities that are specific to services and separated from the USM and SSM (e.g. in a video conference service it can be a video conference session trader).
- *SSSO*: A service substance support object models service-independent capabilities that may support the execution of any application.
- *SF*: A service factory controls the life cycle of the service session objects according to requests from user agents. It may include mechanisms to schedule the activation of a session (e.g. date and time), i.e. reservation. For management purposes it may provide a list of active sessions in a specific domain and clean up some sessions if requested.
- *CSM*: The communication session manager provides the connectivity functionality to the service session manager. It transforms a request of an application, e.g. a stream, into a request for a network connection and requests a connection coordinator (CC) to provide the connection. The CC deals with the network complexity, handling requests to connect several network access points providing a certain bandwidth and a given quality of service. Note that CSM and CC are defined not within the TINA service architecture but within the

connection management architecture (TINA-C document number TB_JJB.005_1.5_94).

*Putting the pieces together*

Figure 5.10 illustrates the relationships of the main computational objects of a typical telecommunication service scenario: a conference with two participants, with the conference data being exchanged between two user applications. Note that this is a simplified representation of computational objects related to service initiation and maintenance. The dynamic creation of objects concerning service and communication sessions using service factories is left out, as are the details of accounting management and subscription management. Furthermore, we do not address connection management, which is the foundation of the establishment of streams.

Good examples of service component interactions are given in Yagi *et al.* (1995). In the next section we illustrate briefly the generic interactions between the identified computational objects for a TINA user invoking a TINA service and for a user being invited into a TINA service (session).

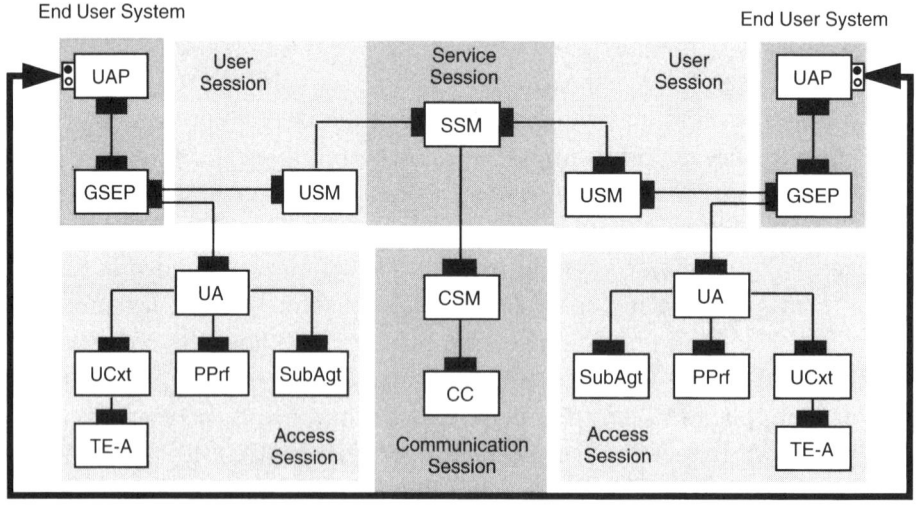

**Figure 5.10**   TINA computational objects involved in a typical service scenario.

### 5.2.4.3 Service component interactions

To help readers to understand how the computational objects of the service architecture interact, some basic service scenarios of a TINA access session will be explained at component level, i.e. the service invocation and the service invitation.

*Invocation of a TINA service*

To begin a TINA-conformant service, an access session has to be invoked (Figure 5.11). This is done by the user interface of a service, the user application (UAP). It contacts a generic session end point (GSEP), asking for an invocation of the specific service (1). The GSEP forwards the request to the initiator's user agent (UA), including authentication data (2). The UA performs an authentication proof to ensure the GSEP's (and the user behind the GSEP) validity and queries a subscription agent (SubAgt) to find out if the user is authorized to use the requested service (3). The SubAgt returns either a non-authorization notification or an authorization. With a positive authorization decision, the SubAgt delivers the service profile containing associated service capabilities and an object reference of a service factory (4) with the capability to create appropriate service session components later on. The UA then asks the usage context (UCxt) for an appropriate terminal associated to the user that satisfies the requirements expressed by the service profile (5). The UCxt consults TE-As to ascertain the capabilities and states of registered terminals needed to perform the selection task and returns the terminal and network access point (6). Having passed the authorization and capability checks, the UA queries the personal profile (PPrf) for the user's constraints and preferences for individual service execution (7). Having received the personal service settings (8), the invitation phase of the access session is complete.

The creation of the service session is done by the UA. It asks the SF (remember that the object reference for the SF came from the SubAgt) to create a user service session manager (USM) and a service session manager (SSM) (9). Whereas the SSM controls the service session, the USM maintains the role of only one participant in the service session. The SF creates the USM and SSM (10), giving them object references for each other (10), and returns these references to the UA (11). The UA relays the result of creation and the reference for the USM to the GSEP (12). The GSEP forwards the USM reference to the UAP (13) and the service session components start interactions (14).

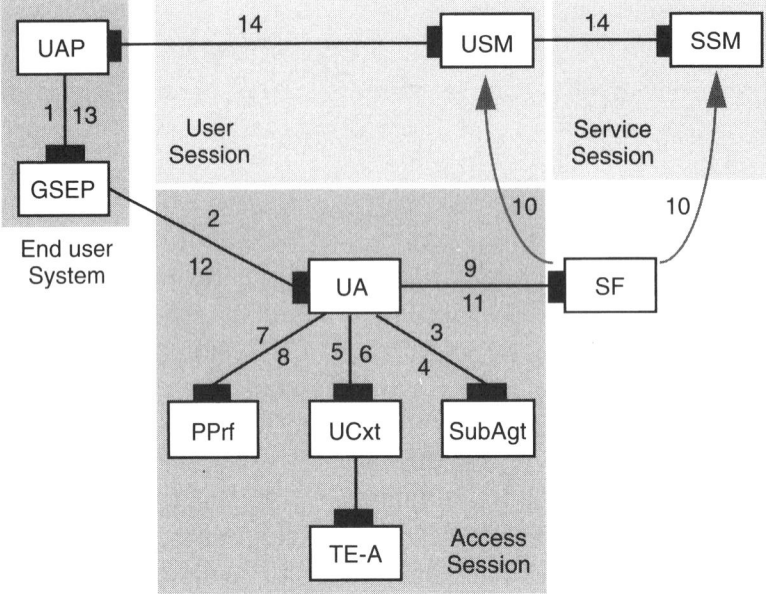

**Figure 5.11** TINA service invocation.

*Service invitation*

Having established a service session, the user application of the session initiator, in this scenario called 'user A', can invite others to join the service session
(Figure 5.12). Whenever user A sends an invitation to another user, the $UAP_A$
relays it to its $GSEP_A$ (1). The $GSEP_A$ passes the invitation to its user service
session manager ($USM_A$) (2), which relays it to the SSM (3). The SSM then
contacts the user agent (UA) of the invited person (4). As in service invocation, the UA first queries the subscription agent to check the subscription
characteristics of the user with respect to the required service (5). If the invited
user is not authorized, the SubAgt notifies a subscription failure for the invited
user and the UA replies to the SSM that the user is not allowed to join this
service (18'). Otherwise, the SubAgt returns a service profile and an object reference of a service factory capable of creating a USM for the specific service (18).
Afterwards, the UA asks the usage context for a terminal at which the user
can be reached (7). The UCxt returns a terminal identifier, a network access
point to the UA (8) leading to an object reference of the terminal dedicated
GSEP. The UA then determines the constraints and preferences to apply an
individual service execution for the invited user. This is done by querying the

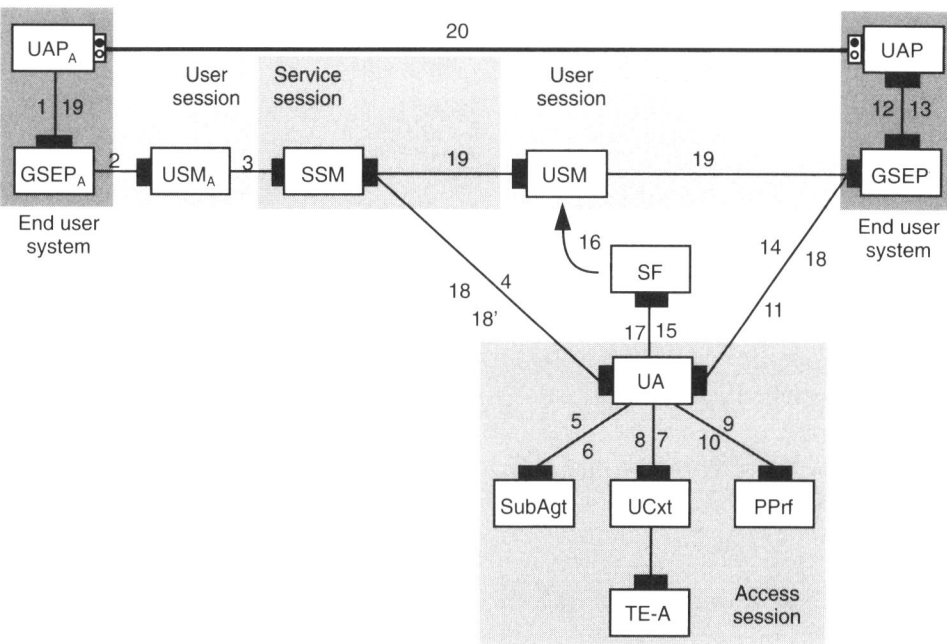

**Figure 5.12** TINA service invitation.

personal profile (9,10). The UA then delivers the invitation to the GSEP (11), which asks the UAP to alert the user (12). If there is no UAP running, the GSEP creates a new one. With the deliverance of the invitation to the user, the basic invitation phase is complete.

As soon as the end user has decided about the invitation, the UAP sends a response to the GSEP (13), which relays it to the UA (14). If the user decided to reject the invitation, the UA indicates an appropriate response to the SSM (18'), which relays it through the USM and the GSEP to the UAP (without numbered equivalents in Figure 5.12).

If the response indicates acceptance, the UA initiates a join phase, asking the service factory (SF) to instantiate a user SSM (15). The SF creates a USM (16), indicates success or failure to the UA, and returns the object reference of the created object (17). The UA then returns the USM object reference to the GSEP and an acceptance notification to the SSM (18). Finally, the USM can be bound to the SSM, so that the service session is extended by a user session (19). The accepted user invitation can be signaled to the $UAP_A$ (19) (through the SSM, USM, and GSEP) and the connection management may establish a stream binding between the user applications for the service, e.g. a video stream for a multimedia conference (20).

## 5.2.5 TINA network architecture

The purpose of the network architecture is to provide a set of generic concepts that describe transport networks in a technology-independent way. The network architecture defines a set of abstractions that the service architecture can work with. At one end it provides a high-level view of network connections to services. At the other end it provides a generic descriptions of elements that can be customized to particular technologies and products.

The concept of *connection graph* is used to provide a service-oriented view of connectivity. A connection graph depicts vertices and ports, with ports connected by lines to represent connectivity. There are two types of connection graphs.

1. In a *logical connection graph* the vertices represent objects, the ports stream interfaces, and the lines streams.
2. In a *physical connection graph* the vertices represent physical nodes, the ports network access points, and the lines connections.

A service can express connectivity requirements by building a logical connection graph (i.e. requesting the establishment of streams between computational objects), and request a CSM to build an appropriate physical connection graph and intranodal bindings. This means that a CSM accepts logical connection graphs as input from a service (i.e. from an SSM) and produces a physical connection graph as output. It is responsible for negotiating with connection management to establish a physical connection graph (i.e. the connections), and to interact with nodes to perform internal network access point/stream interface bindings. The basic principles are defined within the TINA connection management architecture (TINA-C document number TB_JJB.005_1.5_94).

In addition, the network architecture has defined an information specification of transmission and switch technologies, known as the network resource information model (NRIM), from which the technology-dependent aspects have been extracted.

## 5.2.6 TINA management architecture

The management architecture (TINA-C document number TB_GN.010_2.0_94; Berndt *et al.*, 1995) provides a set of generic management principles and concepts for the TINA service, network, and computing architecture, i.e. all the other TINA architectures are strongly influenced by the management architecture principles. In general, the management architecture is based primarily on OSI management and TMN standards. In particular, the management architecture adopts the TMN functional layers (section 4.1), with the service architecture focusing on the service management layer and the network architecture on the network management and network element management layer. The business management layer is not yet considered within TINA. Furthermore, TINA follows the management functional areas defined within OSI management and TMN, namely configuration, fault, performance, accounting, and security management.

Looking at the management aspects covered within the service architecture, the emphasis has been put on the configuration management functional area and specifically on the management of the session types described in section 5.2.4. The session management comprises those functionalities required to activate, modify, suspend, resume, and complete a session plus the functions provided by the session itself. Thus, some computational objects are referred to as 'session managers' within the service architecture. In addition, TINA has elaborated on accounting and subscription management by the definition of generic accounting and subscription computational objects.

Within the network architecture TINA focuses on the management of resources that are within the scope of the network architecture. In this context TINA adopts the five classical management functional areas with one important enhancement. TINA breaks down the traditional 'configuration' management functional area into *'connection management'* and *'resource (configuration) management'*. Resource configuration management comprises installation support, provisioning, and status and control of network resources. Connection management includes the functionalities related to the set-up, management, and releasing of connections.

In particular, connection management as defined in TINA-C document number TB_JJB.005_1.5_94) is a fundamentally new approach to the traditional way of connection control. Connection management functions reside in the network and network element management layers. The CSM, which is considered to be located in the network management layer, provides the connectivity functionality to the services that are considered to be located in the service management layer (Figures 5.9 and 5.10). The CSM transforms the requests from a service, i.e. an SSM, into a request for a network connection by interacting with a connection coordinator. The connection coordinator interacts with other connection management-related computational objects to achieve the connection.

## 5.2.7 Evolution from IN to TINA

The above brief description illustrates that there are fundamental differences between the function-oriented IN architecture and the object-oriented TINA. In contrast to the IN approach of decomposing services into reusable service features and SIBs, within TINA services are modeled by interacting computational objects. Rather than defining functional network elements for the distributed implementation of SIBs as is done in the IN architecture, within TINA the distributed processing environment supports the arbitrary distribution of TINA service components, i.e. the computational objects. Hence, there is no need for specific network nodes with dedicated functionality within TINA. Furthermore, it must be emphasized that the information flows between IN functional entities (via the signaling network) will be replaced within TINA by computational object interactions via operational interfaces (supported by the DPE). Finally, an IN service call, i.e. the established bearer connection, will be modeled as a stream within TINA.

Hence, the interworking between IN and TINA platforms as well as the evolution from current IN standards toward TINA is a challenging task (Brown, 1994; TINA-C document number TP_AJH.001_0.10_94; Magedanz, 1995). Several research activities began in 1995 in order to study this issue closely. One important project in this context is the EURESCOM project P508, 'Evolution, Migration Paths and Interworking with TINA', which is also a TINA-C auxiliary project. (Note that EURESCOM results are usually considered proprietary to the EURESCOM stakeholders. However, it is possible that in the future some project deliverables will be made available for the general public. You should consult EURESCOM headquarters for information on this issue. The address is: EURESCOM GmbH, Schloss Wolfsbrunnenweg 35, D-69118 Heidelberg, Germany.)

When addressing the interworking and migration issues for TINA, it should be kept in mind that TINA was primarily designed to support future multimedia multiparty services. Since today's IN platforms provide a mature foundation for advanced telephony services on top of the PSTN and the ISDN and on the other hand represent a fundamental investment for network operators, there is no big motivation for replacing these IN platforms with emerging TINA platforms in the short term. However, with increasing demand for multimedia applications on top of emerging broadband networks, i.e. B-ISDN, it will become increasingly important to study possible interworking options between IN and TINA systems as well as a migration path from IN toward TINA. Below we outline briefly the basic issues.

Interworking between IN and TINA systems becomes necessary in cases where multiple network operators provide internetwork IN services, such as a pan-European VPN service, and where these operators deploy IN and TINA platforms. Hence a corresponding interworking function (IWF) is required which enables interactions between an IN SCF/SDF within one operator's domain and a set of interacting TINA service components on top of a DPE within another operator's domain. Thus, basically, INAP operations have to be mapped onto operations provided by the TINA objects and vice versa. This scenario can be compared with IN interworking addressed in section 3.2.2.1 on page 104.

Migration or evolution of a pure IN platform toward TINA means a step by step 'TINArization' of the IN platform elements, i.e. specific IN functional entities will be replaced by means of appropriate TINA service components. Such a replacement of IN functional entities has to start with those IN elements, that are not yet implemented or deployed in small numbers. This means that the SSFs already deployed on a global basis in the exchanges, representing the biggest amount of IN related investment, would not be replaced until it is necessary. This allows one to keep the existing BCSM and the INAP interfaces in operation. It seems reasonable to start with a TINArization of the SDF and subsequently TINArize the SCF (Figure 5.13).

Within the first migration step IN service data is TINArized. Thus functions envisaged for the IN SDF will be provided by TINA service components defined within the service architecture, i.e. the personal profile, usage context, service profile, subscription agent, etc. This approach allows one to take advantage of TINA data modelling concepts and data distribution transparency provided by the DPE. Consequently an IWF is required, enabling IN SCF access to service operational data and service management data within the TINA platform. The IWF receives INAP messages from the IN SCF and sends requests to TINA computational objects implementing the SDF functionality. Furthermore the IWF translates the results from the TINA objects back into INAP messages for the SCF. Notice that as an option within this evolution step also the service management function (SMF) may also be TINArized. This would ease IN service data management to a large extent.

The next migration step toward TINA encompasses the TINArization of IN service logic. Hence the SCF will be realized by corresponding TINA computational objects, e.g. the user agent, service session manager, communication manager, etc. This means that within this scenario both IN service logic and service data are realized by TINA service components defined within the service architecture. This allows one to take full advantage of object-oriented service modelling, in particular software reuse. Only the SSF remains within the IN domain, which requires another IWF, which maps the INAP invocations from an SSF into requests to TINA computational objects implementing the SCF functionality and translates the results from the TINA objects back into INAP messages for the SSF. (Note that the SRF still exists, but it is considered to be of less importance.)

The final evolution step from IN toward TINA encompasses an adaption of the BCSM and the introduction of DPEs within the exchanges. By this approach the INAP interface and the IWF are no longer required and the exchanges can directly invoke the operations provided by the TINA computational objects via the DPE. Full advantage can be taken of the distribution transparency provided by the DPE. Note that the IN has vanished.

Whether the TINA architecture can replace the IN architecture in the long term will depend on the success of the migration studies and the performance of first TINA platform prototype implementations. Nevertheless, it can be assumed that TINA concepts will have a strong influence on future IN capability sets.

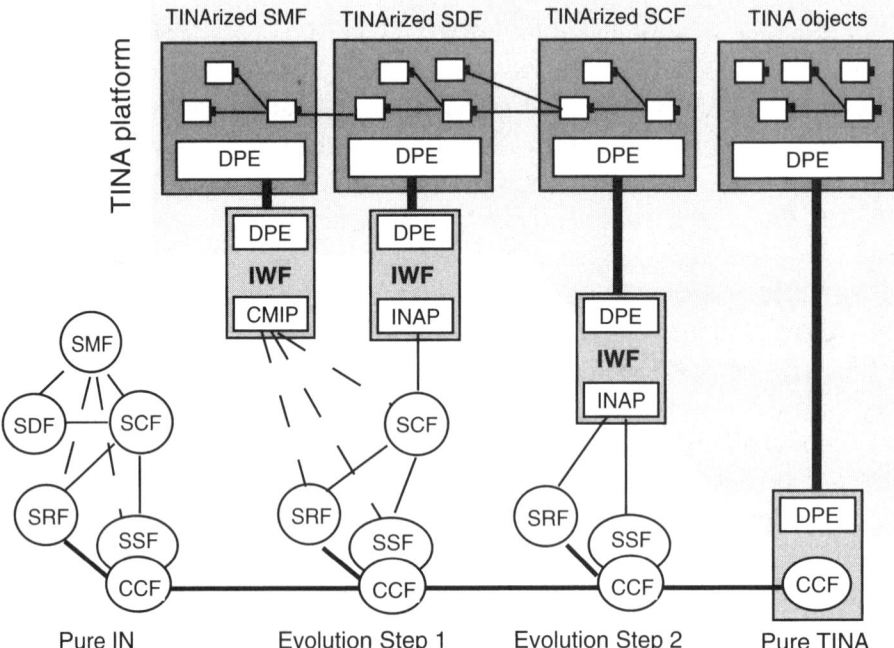

**Figure 5.13** IN evolution toward TINA.

In summary, the shift from function-oriented modeling approaches toward object-oriented modeling concepts and the increasing use of software 'agents' in advanced telecommunications architectures will have important effects on the telecommunications environment in general (for further details see Reinhardt, 1994; Bagely et al. 1995; Magedanz *et al.*, 1996). In particular, significant impacts on the existing signaling concepts and protocols can be expected, resulting in the definition of new network/service elements, such as terminal agents, user agents, and communication managers. This may result in a replacement of the traditional IN service switching points and service control points (Lauer *et al.*, 1995) in the medium-term.

# 6 Summary

This book has presented an overview of the intelligent network as it is understood today by the international standards bodies. We have illustrated the motivation and the technological basis for the IN concept. We have presented the historical evolution from the traditional telecommunications network environment toward an IN-structured network, focusing on the main goals and principles of the IN concept. The main focus of this book is on the IN conceptual model, representing the general framework for IN development today. In this context we have looked particularly at the international IN standards, which provide the foundation for current IN implementation. Additionally, we have addressed the interrelationships between the IN and other emerging concepts in the telecommunications environment, such as TMN, B-ISDN, and mobile communication systems. Furthermore, we looked briefly at the future evolution of the IN, driven by the convergence of computing and telecommunications and the effects of object orientation on telecommunication in particular. In this context we have focused on the new telecommunications information networking architecture, which is currently gaining momentum.

We have tried to keep the description as short and simple as possible, but the field of IN is complex and related to many other technologies. It is very difficult to maintain simplicity while aiming to provide the basis for deeper investigations. Our emphasis was to provide the main ideas, concepts, and terminology in the field of IN, so that interested readers will be prepared to study the details in the referenced literature. As such, this book can be seen a kind of navigation guide through available IN standards and IN publications.

Today INs are deployed all around the world. Many IN services will be offered on a national basis in many countries based on proprietary IN platforms. The first standard-based IN platforms are likely to appear in the 1995–96 time frame. In this book we have excluded a survey of existing IN platforms and services offered or planned in the different countries. We believe that such surveys are beyond the scope of an IN tutorial, as the rapid progress in IN development makes market surveys out of date shortly after publication.

# Appendix A
# CS-1 services, service features, SIBs

## A.1 CS-1 services

The descriptions provided below of CS-1 benchmark IN services are compiled from various sources (ITU Recommendation series Q.121x; ETSI TCR-TR NA-60106). The service descriptions mirror the user's view of the service as it is given by the IN service plane of the IN conceptual model. Note that these benchmark services serve only as a requirement for the definition of IN SIBs and their realization within the distributed IN architecture. The standardization of IN services and service features is outside the scope of IN standardization.

- *Abbreviated dialing* (ABD): This service allows subscribers to call others by dialing an abbreviated number.
- *Account card calling* (ACC): This service enables subscribers to use a telephone card to make telephone calls from any card-reading telephone. For identification the user receives an access code and a personal identification number (PIN). Thus, the caller dials the access code, enters the PIN and draws the card through the reader. The card content defines a domestic or business account number to which charges for the call are debited.
- *Automatic alternative billing* (AAB): (1) This service enables a user to be charged for a call by the user's account from any telephone, i.e.

the call charge does not refer either to the calling line or to the called line. An account code and PIN are allocated to a service user and validated if the user invokes a call. (2) This service allows the user (as calling party) to ask another user (as called party) to receive the call at her or his expense. A call could consist of two steps: after the calling party is invited to record a brief message for the called party, the latter is alerted and asked to accept the call and be charged for it.

- *Call completion to busy subscriber* (CCBS): Calling a busy destination, the user is informed when the busy line becomes free without making a new call.
- *Call distribution* (CD): The incoming calls of the subscriber may be routed to different destinations according to subscriber-specific rules. In accordance with the agreed control capabilities, the subscriber is able to change the rules in real time. If the chosen destination is busy the call may be routed automatically to an alternative location.
- *Call forwarding* (CF): By invoking this service via subscriber control capabilities the user may forward incoming calls to another telephone number.
- *Call rerouting distribution* (CRD): Encountering a trigger condition (busy, queue overload, etc.) the incoming calls of the subscriber are rerouted to a predefined destination.
- *Conference calling* (CON): This service supports the connection of multiple parties in a single conversation.
- *Credit card calling* (CCC): The subscriber may call from any normal access interface to any destination number and is charged to an account specified by the CCC number. This account may also be a bank card account.
- *Destination call routing* (DCR): The subscriber's incoming calls are routed to different destinations according to various conditions (time of day, area of call origination, etc.). Additionally, the subscriber may obtain statistics related to the calls.
- *Follow-me diversion* (FMD): This service provides subscribers with facilities to redirect calls from their primary telephone number to other locations (including terminal access) via customer control capabilities.
- *Freephone* (FPH): The subscriber of the Freephone service is able to reverse charging by accepting charges for the incoming calls. This means that calling users do not have to pay the call charges when calling a Freephone number. Freephone subscribers are usually but not necessarily large companies using the service for promotion purposes.

- *Malicious call identification* (MCI): This service allows the subscriber to control the logging (record) of incoming calls.
- *Mass calling* (MAS): The MAS service provides capabilities for instantaneous, high-volume traffic routed to one or more destinations depending on specific conditions. The subscriber may obtain statistics concerning the calls (e.g. number of calls).
- *Originating call screening* (OCS): The subscriber is able to screen outgoing calls in accordance with a screening list that either restricts or allows them. The user in turn may override the restriction on a per call basis by means of an authorization code (identity code or PIN).
- *Premium rate* (PRM): This service provides the subscriber (usually an information service provider) with a premium rate number. A user who calls this number is charged at a special rate for both the call and the information/service obtained by the call. The network operator collects the revenue and shares it with the service provider.
- *Security screening* (SEC): Security screening is intended to hinder unauthorized access to the subscriber's network, systems, or applications. Invalid attempts may be recorded.
- *Selective call forwarding on busy/don't answer* (SCF): The user's incoming calls can be redirected according to a screening list, regardless of the user's line status. If the user's line is busy or the user does not answer (within $x$ rings/seconds) the incoming call is redirected to another number.
- *Split charging* (SPL): This service allows charge splitting: the calling as well as the called party is charged for a part of the call.
- *Televoting* (VOT): This service enables the subscriber to perform phone voting by allocating one or more temporary numbers. The service may provide facilities to play announcements acknowledging the call, to count the calls, and to supply the total number of calls. The temporary number(s) are reallocated and the calls may be charged at varying rates.
- *Terminating call screening* (TCS): The subscriber can construct a screening list to specify whether incoming calls are restricted or allowed.
- *Universal access number* (UAN): The terminating lines of the subscriber at different locations can be reached via a unique number. Based on the area from which the incoming call originates, the subscriber may specify a number to which the call is routed. The service may provide capabilities for statistics concerning the incoming calls.

- *Universal personal telecommunication* (UPT): UPT provides the subscriber with 'mobility'. Based on a unique personal telecommunication number (PTN), the subscriber may use (initiate and receive) any telecommunication service across multiple networks and at any user network access (fixed, movable, or mobile). Irrespective of the location and limited only by the terminal and network capabilities, the PTN is translated to the current destination number of the subscriber.
- *User-defined routing* (UDR): This service allows the subscriber to determine the routing for outgoing calls in agreement with a routing preference list.
- *Virtual private network* (VPN): This service provides private network capabilities by using public network resources. The subscriber's lines, connected to different network switches, constitute a virtual private network including capabilities such as private numbering plan (PNP) and call transfer (TRA).

## A.2 CS-1 service features

- *Abbreviated dialing* (ABD): Facilitates the definition of abbreviated dialing numbers representing the actual dialing number, i.e. a short digit sequence is a complete dialing sequence for a private or public numbering plan.
- *Automatic call back* (ACB): Allows a called party automatically to call back the calling party of the last call directed to the called party.
- *Attendant* (ATT): Permits VPN users requiring VPN service information to access an attendant of the VPN. By dialing a special access code the user reaches this attendant, who provides the required information.
- *Authentication* (AUTC): This service feature allows the verification of a user who needs to perform certain operations in an IN.
- *Authorization code* (AUTZ): Empowers a user to override calling restrictions of the terminal from which the call is made. Different sets of calling privileges can be assigned to different authorization codes and a given authorization code can be shared by multiple users.
- *Call distribution* (CD): Allows the served user to specify the percentage of calls that are to be distributed among two or more destinations. Other criteria may also apply to the distribution of calls to each destination.

- *Call forwarding (unconditional)* (CF): The user's incoming calls can be addressed to another number, no matter what the called party line status may be.
- *Call forwarding (conditional)* (CFC): Allows the called user to forward particular calls if the called user is busy or does not answer within a specific number of rings.
- *Call gapping* (GAP): Allows the service provider automatically to restrict the number of calls routed to the subscriber.
- *Call hold with announcement* (CHA): Allows a subscriber to place a call on hold with options to play music or customized announcements to the held party.
- *Call limiting* (LIM): Allows the called party to specify a maximum number of simultaneous calls to a served user´s destination. As an option the threshold may be real-time managed by the subscriber.
- *Call logging* (LOG): Provides means for keeping a record of calls to a specified telephone number.
- *Call queuing* (QUE): The served user's incoming calls are placed in a queue if the line is busy and connected as soon as the line becomes free. When entering the queue, the caller hears an initial announcement informing the caller that the call will be answered when a line becomes available.
- *Call transfer* (TRA): Allows a subscriber to place a call on hold and transfer the call to another location.
- *Call waiting* (CW): Allows the called party to receive a notification that another party is trying to reach his number while he or she is busy talking to another calling party.
- *Closed user group* (CUG): A VPN user can be assigned to a group of VPN users authorized to make and/or receive calls only within the group. A user can belong to more than one CUG. In this way, a CUG can be defined so that certain users are allowed either to make calls outside the CUG, or to receive calls from outside the CUG, or both.
- *Consultation calling* (COC): Allows a subscriber to place a call on hold for initiating a new call for consultation.
- *Customer profile management* (CPM): The subscriber is provided with the capabilities to manage her or his service profile in real-time, i.e. specify call distribution, terminate destination, play announcements, etc.
- *Customer recorded announcement* (CRA): Allows a call to be completed to a (customized) terminating announcement instead of a subscriber

line. The served user may define different announcements for unsuccessful call completions depending on the reasons (e.g. caller outside business hours, all lines are busy).

- *Customized ringing* (CRG): Allows a subscriber to allocate a distinctive number of rings to a list of calling parties.
- *Destination user prompter* (DUP): Enables the prompting of the called party with a specific announcement. Such an announcement may ask the called party to enter an extra number, e.g. through dual tone multi-frequency (DTMF), or a voice instruction that can be used by the service logic to continue the processing of the call.
- *Follow-me diversion* (FMD): Allows a user to register for incoming calls to any terminal access. When registered, all incoming calls to the user will be presented to this terminal access. A registration for incoming calls will cancel any previous registration. Several users may register for incoming calls to the same terminal access simultaneously. The user may also explicitly deregister for incoming calls.
- *Mass calling* (MAS): Allows processing of huge numbers of incoming calls generated by broadcast advertising or games.
- *Meet-me conference* (MMC): Allows a user to reserve a conference resource for making a multiparty call, indicating the date, time, and conference duration. At the specified date and time, each participant in the conference has to dial a specific number, which has been assigned to the reserved conference resource.
- *Multiway calling* (MWC): Allows a user to establish multiple, simultaneous telephone calls with other parties.
- *Off-net access* (OFA): Allows a VPN user to access her or his VPN from any non-VPN station in the PSTN by using a PIN. Different sets of calling privileges can be assigned to different PINs and a given PIN can be shared by multiple users.
- *Off-net calling* (ONC): Allows the user to call outside the VPN network. Calls from one VPN to another are also considered to be off-net.
- *One number* (ONE): Allows a subscriber with two or more terminating lines in any number of locations to have a single telephone number. The subscriber may specify which calls are to be terminated on which terminating lines based on the area from which the calls originate.
- *Originating call screening* (OCS): Allows the served user to bar calls from certain areas based on the district code of the area from which the call originates.

- *Origin-dependent routing* (ODR): Enables the subscriber to accept or reject a call and in case of acceptance to route this call according to the calling party's geographical location. This service feature allows the served user to specify the destination installation(s) according to the geographical area from which the call was originated.
- *Originating user prompter* (OUP): Enables the prompting of the calling party with a specific announcement. Such an announcement may ask the calling party to enter an extra number (e.g. through DTMF) or a voice instruction that can be used by the service logic for continuing to process the call.
- *Personal numbering* (PN): Supports a UPT number that uniquely identifies each UPT user and is used by the caller to reach that UPT user. A UPT user may have more than one UPT number for different applications (e.g. a business UPT number for business calls and a private UPT number for private calls), however a UPT user will have only one UPT number per charging account.
- *Premium charging* (PRMC): Allows for the payback of part of the cost of a call to a called party that is considered to be a value-added service provider.
- *Private numbering plan* (PNP): Empowers VPN subscribers to define a private numbering plan within their private network, which is separate from the public numbering plan.
- *Reverse charging* (REVC): Allows the service subscriber to be charged for the entire cost of a call.
- *Split charging* (SPLC): Allows the separation of charges for a specific call; the calling and called party are each charged for part of the call costs.
- *Terminating call screening* (TCS): Allows the user to screen calls based on terminating telephone number dialed.
- *Time-dependent routing* (TDR): Allows the served user to apply different call treatments based on time of day, day of week, day of year, holiday, etc.

# A.3 Mapping of CS-1 services and service features

The following tables provide an overview of core service features (closed circles) present in each CS-1 IN service. In addition, the tables indicate which optional service features (open circles) could be added to enhance the basic service functionalities.

| Services \ Service Features | Abbreviated Dialling | Automatic Call Back | Attendant | Authentication | Authorization Code | Call Distribution | Call Forwarding (Uncond.) | Call Forwarding Cond. | Call Gapping | Call Hold with Announce. | Call Limiting | Call Logging | Call Queuing | Call Transfer | Call Waiting | Closed User Group | Consultation Calling | Customer Profile Mgt. | Customer Recorded Ann. |
|---|---|---|---|---|---|---|---|---|---|---|---|---|---|---|---|---|---|---|---|
| Abbreviated Dialling | ● | | | | | | | | | | | ○ | | | | | | ○ | |
| Account Card Calling | ● | | | | ● | | | | | | | ○ | | | | | | | |
| Automatic Alternative Billing | | | | | ● | | | | | | | ○ | | | | | | | |
| Call Completion to Busy Subs. | | ● | | | | | | | | | | ○ | | | | | ○ | | |
| Call Distribution | | | | | | ● | | | | | | ○ | | | | | | ○ | |
| Call Forwarding | | | | | | | ● | | | | | ○ | | | | | | ○ | |
| Call Re-routing Distribution | | | | | | | | ○ | | | ○ | ○ | ○ | | | | | ○ | ○ |
| Conference Calling | | | | | | | | | | | | ○ | | | | ○ | ○ | | |
| Credit Card Calling | ○ | | | | ● | | | | | | | ○ | | | | | | | |
| Destination Call Routing | | | | | | ● | | | | | | ○ | | | | | | ○ | |
| Follow-Me-Diversion | | | | | | | | | | | | ○ | | | | | | ○ | |
| Freephone | | | ○ | | ○ | | ○ | ○ | | | ○ | ○ | ○ | | | | | ○ | ○ |
| Malicious Call Identification | | | | | | | | | | | | ○ | | | | | | | |
| Mass Calling | | | | | ○ | | | ○ | | | | ○ | ○ | | | | | ○ | ○ |
| Originating Call Screening | | | | | | | | | | | | ○ | | | | | | ○ | |
| Premium Rate | | | | | ○ | | ○ | ○ | | | ○ | ○ | ○ | | | | | ○ | ○ |
| Security Screening | | | | ● | | | | | | | | ○ | | | | | | ○ | |
| Selective Call Forwarding | | | | | | | ● | | | | | ○ | | | | | | ○ | |
| Split Charging | | | | | ○ | | ○ | ○ | | | ○ | ○ | ○ | | | | | ○ | ○ |
| Televoting | | | | | ○ | | | ○ | | | ○ | ○ | ○ | | | | | ○ | ○ |
| Terminating Call Screening | | | | | | | | | | | | ○ | | | | | | ○ | |
| Universal Access Number | | | | | ○ | | ○ | ○ | | | ○ | ○ | ○ | | | | | ○ | ○ |
| Universal Personal Telecom. | | | | | ● | | | | | | | ○ | | | | | | ○ | ○ |
| User-Defined Routing | | | | | | | | | | | | ○ | | | | | | ○ | |
| Virtual Private Network | ○ | | ○ | ○ | ○ | ○ | | | | ○ | | ○ | ○ | ○ | | ○ | ○ | ○ | ○ |

● Core Feature    ○ Optional Feature

| Services \ Service Features | Customized Ringing | Destinating User Prompter | Follow-Me-Diversion | Mass Calling | Meet-me Conference | Multiway Calling | Off-Net Access | Off-Net Calling | One Number | Originating Call Screening | Origin Dependent Routing | Originating User Prompter | Personal Numbering | Premium Charge | Private Numbering Plan | Reverse Charging | Split Charging | Terminating Call Screening | Time Dependent Routing |
|---|---|---|---|---|---|---|---|---|---|---|---|---|---|---|---|---|---|---|---|
| Abbreviated Dialling | | ○ | | | | | | | | | | | | | | | | | |
| Account Card Calling | | | | | | | | | | | | ● | | | | | | | |
| Automatic Alternative Billing | | | | | | | | | | | | ● | | | | | | | |
| Call Completion to Busy Subs. | | | | | | | | | | | | | | | | | | | |
| Call Distribution | | | | | | | | | ● | | ○ | | | | | | | | ○ |
| Call Forwarding | | | | | | | | | | | | | | | | | | | |
| Call Re-routing Distribution | | | | | | | | | ● | | | | | | | | | | |
| Conference Calling | | | | | ○ | ● | | | | | | | | | | | | | |
| Credit Card Calling | | | | | | | | | | | | ● | | | | | | | |
| Destination Call Routing | | | | | | | | | | | ○ | | | | | | | | ○ |
| Follow-Me-Diversion | | | ● | | | | | | | | | | | | | | | | |
| Freephone | ○ | ○ | | ○ | | | | | ● | ○ | ○ | ○ | | | | ● | | | ○ |
| Malicious Call Identification | | | | | | | | | | ● | | | | | | | | | |
| Mass Calling | | | ● | | | | | | | ○ | ○ | ○ | | | | | | | ○ |
| Originating Call Screening | | | | | | | | | | ● | | | | | | | | | |
| Premium Rate | ○ | | | | | | | | ● | ○ | ○ | ○ | | ● | | | | | ○ |
| Security Screening | | | | | | | | | | | | | | | | | | | |
| Selective Call Forwarding | | | | | | | | | | | | | | | | | | | |
| Split Charging | ○ | ○ | | | | | | | ● | ○ | ○ | | | | | | ● | | |
| Televoting | | | ○ | | | | | | | ○ | ○ | ○ | | | | | | | ○ |
| Terminating Call Screening | | | | | | | | | | | | | | | | | | ● | |
| Universal Access Number | ○ | | | | | | | | ● | ○ | ○ | ○ | | | | | | | ○ |
| Universal Personal Telecom. | | ○ | ● | | | | | | | | | ○ | ● | | | ● | | | ○ |
| User-Defined Routing | | | | | | | | | | | ○ | | | | | | | | ○ |
| Virtual Private Network | ○ | | ○ | | | | ○ | ○ | | | | ○ | | | ● | | | | ○ |

● Core Feature     ○ Optional Feature

# A.4 Mapping of service features onto SIBs

| Service Features \ SIBs | Authenticate | Algorithim | Charge | Compare | Distribution | Limit | Call Log Inform. | Queue | Screen | Service Data Mgt. | Status Notification | Translate | User Interaction | Verify |
|---|---|---|---|---|---|---|---|---|---|---|---|---|---|---|
| Abbreviated Dialling | | | | | | | | | | | | ● | | |
| Automatic Call Back | | | | | | | | | | | ● | | | |
| Attendant | | | | ● | | | | | | | | | ● | |
| Authentication | ● | | | | | | | | | | | | ● | ● |
| Authorization Code | ● | | | | | | | | | | | | ● | ● |
| Call Distribution | | ● | | | ● | | | | | | | | | |
| Call Forwarding (Uncond.) | | | | | | | | | | ● | | ● | | ● |
| Call Forwarding Cond. | | | | | | | | | | ● | ● | ● | | ● |
| Call Gapping | | | | | | ● | | | | ● | | | | |
| Call Hold with Announce. | | | | | | | | | | ● | | | ● | |
| Call Limiting | | | | | | ● | | | | ● | ● | | | |
| Call Logging | | | | | | | ● | | | ● | | | | |
| Call Queuing | | | | | | ● | | ● | | ● | | | | |
| Call Transfer | | | | | ● | | | | | | | | | |
| Call Waiting | | | | | | | | | | ● | | | ● | |
| Closed User Group | | | | ● | | | | | | | | | | |
| Consultation Calling | | | | | | | | ● | | ● | | | ● | |
| Customer Profile Mgt. | | | | | | | | | | | | | | |
| Customer Recorded Ann. | ● | | | | | | | | | ● | | | ● | ● |
| Customized Ringing | | | ● | | | | | | | ● | | | | |
| Destinating User Prompter | | | ● | | | | | | | | | | ● | ● |
| Follow-Me-Diversion | | | | | | | | | | ● | | ● | | |
| Mass Calling | | ● | | | ● | ● | | | | | | | ● | |
| Meet-me Conference | ● | | | | ● | | | | | | | | ● | ● |
| Multiway Calling | ● | | | | | | | ● | | ● | | | ● | |
| Off-Net Access | | | | | | | | | | | | | ● | ● |
| Off-Net Calling | | | | | | | | | | | | | ● | ● |
| One Number | | ● | | | ● | | | | | ● | ● | | | |
| Originating Call Screening | | | ● | | | | | | ● | | | | | |
| Origin Dependent Routing | | | | | ● | | ● | | ● | | ● | | | |
| Originating User Prompter | | | | | | | | | | | | | ● | ● |
| Personal Numbering | ● | | | | | | | | | ● | | ● | ● | ● |
| Premium Charging | | | ● | | | | ● | | | ● | | | | |
| Private Numbering Plan | | | | | | | | | | ● | | ● | | |
| Reverse Charging | | | ● | | | | ● | | | ● | | | | |
| Split Charging | | | ● | | | | ● | | | ● | | | | |
| Terminating Call Screening | | | | ● | | | | | ● | | | | | |
| Time Dependent Routing | | | | | ● | | | ● | | ● | ● | ● | | |

# Appendix B
# CS-2 service categories and service features

At the time of writing there exists no complete and consistent list of CS-2 services and service features. However, we think that it is interesting to take a look of the potential CS-2 services and service features. Therefore we provide some examples for possible CS-2 services and service features. For more accurate information readers are referred to the recent ITU Draft Recommendations Q.1221 and Q.1229 and ETSI DTR NA-60902.

## B.1 CS-2 telecommunications services and service features

The envisaged telecommunications services and service features included in IN CS-2 represent a superset of those included in IN CS-1. This means CS-2 benchmark services include the CS-1 services. CS-2 services fall primarily into the category of 'single ended', 'single point of control' services (implicitly, single medium) referred to as Type A services. Nevertheless, CS-2 gives some high level guidelines for supporting new individual telecommunication services, such as mobility services and broadband services which require Type B service support.

The CS-2 benchmark telecommunications services and service features may be grouped as follows:

- internetworking services and features,
- call party handling services and features,
- personal mobility services and features, i.e. UPT,
- terminal mobility services and features, i.e. CTM and UMTS, and
- other services and service features, including multimedia and broadband services.

In the following we provide some examples for each service category. Note that the names of the services and services features as well as their description may be subject of changes.

# B.1.1 Internetworking services and service features

- *Internetwork Freephone* allows the served user having one or more installations to be reached from a specific network other than his/her home network with a Freephone number, and to be charged for this kind of call.
- *Internetwork Premium Rate* provides two-way interactive communication between callers in one network and service/information providers in another network. The calling party is charged with a premium rate for this kind of call.
- *Internetwork Mass Calling* enables to accommodate large volumes of simultaneous calls to a single destination number in another network. It can provide one-way, non-interactive communication between each caller in a given network and a service/information provider in another network. Using this service, the network operator can temporarily allocate a single service number to the served user.
- *Internetwork Televoting* allows a service/information provider in one network to conduct voting or polling over the phone. The caller in another network votes by placing a call to a specific number corresponding to a voting/polling choice. The service provides communication between each caller in a given network and a service/information provider in another network. The service/information provider receives a report of the number of calls to each number.
- *Global Virtual Network Service* is a global switched VPN service supported by multiple networks (e.g., offered to customers over PSTN and/or ISDN).

- *International Telecom Charge Call* allows the holders of a telecommunication charge card to make use of a variety of telecommunications services provided by the card acceptor (e.g. the visited network) and have the charges billed to the customer's account number by the card issuer (home network).

The following service features may be involved in the realization of these services:

- *Internetwork Service Identification* permits the receiving network, in an inter-network call, to receive from the originating network an indication of the service used in the received call.
- *Internetwork Rate Indicator – Forward* encompasses the ability to provide across networks, in the forward direction, an indication of the rate either being charged or to be charged for the presented call.
- *Internetwork Rate Indicator – Backward* encompasses the ability to provide across networks, in the backward direction, an indication of the rate either being charged or to be charged for the received call.
- *Real Time Flexible Rating* encompasses the ability to vary in real time, for a given call, the billing rate, or the party being charged. This could be done at subscriber's direction, during a call or during call set up.
- *Originating Carrier Identification* permits the receiving network, in an inter-network call, to receive an indication identifying the 'originating carrier' (i.e., the originating network/Network Operator).
- *Terminating Carrier Identification* permits the receiving network, in an inter-network call, to receive from the originating network, an indication identifying the network where the call is destined to, or 'terminating carrier' (i.e., the terminating network/Network Operator).
- *Internetwork Service profile interrogation* enables a user to interrogate (read only) the current contents of the user's service profile. The profile information could include information such as telecommunications services subscribed to, default parameters, activated supplementary services, current registrations for incoming and outgoing calls, etc.
- *Internetwork Service profile modification* enables a user to modify (read and write) the appropriate user's service profile parameters that are allowed to be modified. Such parameters could be actication or de-activation of supplementary services, various default parameters, PIN code, etc.

- *Internetwork service profile transfer* allows service profile information to be transferred to other service profile storage locations in other networks. Information which is transferred to each visited network may depend on the contract with home network. The information updated in a visited network is transferred to the home network and vice versa.
- *Charge Card Validation* provides the International Telecom Charge Call service an authentication feature to compare user-side information, provided to the visited network, with information stored in the home network.
- *Call Disposition* provides the International Telecom Charge Call service with verification that a (calling) card has enough spare credit (e.g. the card usage value has not been exceeded) to give the permission to make the call.

## B.1.2 Call party handling services and service features

- *Completion of Call to Busy Subscriber* enables a calling user encountering a busy destination to have the call completed when the busy destination becomes not busy, without having to make a new call attempt.
- *Conference Calling* enables a group of users to be connected into a multi-party call.
- *Call Hold* allows a user to place a call on hold and play an announcement to the held party, and to initiate a new call. The user can subsequently resume participation in the original call.
- *Call Transfer* allows a user to place a call party on hold and to enter a new destination number (optionally service logic can provide the destination number). Upon successful call set up, the subscriber is released and the held party is connected to the new destination in a two party active call.
- *Call Waiting* allows a user to notify a subscriber of the occurrence of a call termination attempt, while that subscriber is participating in an active call. Upon the subscriber's request, the network is able to place on hold the call party participating in a previous active call, and allow the subscriber to accept the incoming call. The subscriber is then associated with both calls and able to toggle between the two calls, causing the other call parties to toggle between hold and active conditions.

The following service features may be involved in the realization of these services:

- *Automatic Call Back* allows the called party to automatically call back the calling party of the last call directed to the called party.
- *Call Hold* allows a user to interrupt his/her connection to an existing call, without releasing that call. Some of the resources which were dedicated to that call (e.g. bearer capability) become available for other uses.
- *Call Retrieve* allows a user to re-establish his/her connection to a call previously placed on hold.
- *Call Transfer* allows a user who is a party in two separate calls, to cause the other two parties to those calls to be connected to each other, releasing him/her from both.
- *Call Toggle* is applicable to a user who has one active call and one on hold. It allows him/her repeatedly to select the currently held party as the new connection, the previously connected party being automatically put on hold.
- *Call Waiting* informs a user already engaged in a call that another party is trying to establish a connection to him/her. Signalling means are provided to enable the user to instruct the network as to what further action it should take.
- *Meet-Me Conference* allows the user to reserve a conference resource for making a multi party call, indicating the date, time, and conference duration. At the specified date and time, each participant in the conference has to dial a designated number which has been assigned to the reserved conference resource, in order to have access to that resource, and therefore, the conference.
- *Multi-way Calling* allows the user to establish multiple, simultaneous telephone calls with other parties.
- *Call Pick-up* enables a user to associate a call request to an already alerting call. The alerting call awaits answer while the user originating call pick-up signals to the network a desire to connect to the alerting call. The network then connects the call parties.

# B.1.3 Personal mobility services and service features

CS-2 defines no personal mobility service but refers to UPT. However, the following list of service features is given for personal mobility support. (Note that some service features required for UPT are arleady given in other subsections, such as service indication or internetwork service profile interrogation.)

- *User Authentication* confirms the identity of a user with the Network and the identity of the network with a user.
- *User Registration* enables a user to register on a terminal access for the purpose of receiving or placing calls. This may be accomplished either from the terminal where the user wishes to register or from another terminal (remote registration).
- *Outgoing Call Registration* allows a user from the current terminal address to register for outgoing calls to be made from that terminal address.
- *Secure Answering* enables the service subscriber / user that incoming calls cannot be answered unless the answering party first successfully authenticates himself as the wanted subscriber.
- *Reset of UPT registration for incoming calls* enables any person (the third party), even if not an UPT user, to reset any UPT registrations for incoming calls on the third party's terminal.
- *Follow-on* enables a user to make a series of service requests without repeated identification and authentication process before each service feature request.
- *Flexible (Call) Origination Authorisation* can take effect immediately prior to the time that an IN capable switch would authorise call origination, during the call set up process. A customised algorithm, provided by the network provider or the subscriber, can then determine whether or not the call should be originated.
- *Flexible (Call) Termination Authorisation* can take effect immediately prior to the time that an IN capable switch would authorise call termination, during the call set up process. A customised algorithm, provided by the network provider or the subscriber, can then determine whether or not the call should be authorised.
- *Provision of Stored Messages* informs the subscriber of the service and sends the voice messages which were stored before, when the service subscriber registers at a location.

- *Multiple terminal address registration* enables a user to be registered for incoming calls on more than one terminal.
- *Intended Recipient Identity Presentation* allows identification at the receiving terminal of the intended recipient of an incoming call.
- *Blocking/unblocking of incoming calls* enables any person, even if not a UPT user, to block and unblock calls incoming to UPT users currently registered on the third party's terminal.

# B.1.4 Terminal mobility services and service features

CS-2 defines no terminal mobility service but refers to CTM and UMTS/FPLMTS. However, the following list of service features is given for terminal mobility support. (Note that some service features required for UMTS/FPLMTS are not under this section as they are of more general nature, such as inter-network service profile interrogation, mobile call origination or mobile user call termination.)

- *Terminal Authentication* is initiated within the mobility processes of Location Management (i.e., terminal location registration), Call Origination, Call Delivery and at other times as initiated by the network or terminal. In support of security, the feature ensures the validity of the terminal and the network.
- *Handover* enables a mobile terminal to change network access areas/points within a network or to an other network, while maintaining the call(s) and/or signalling relationship(s). Active services should be maintained, within the limits imposed by available radio and network resources. In the case of a lack of resources, active services could be modified (e.g. fall back to a lower grade of service quality) or interrupted.
- *Terminal location registration* is used when terminals notify the system on their location. This is a feature by which the location area information of a terminal is registered with the network.
- *Terminal Attach/Detach* The attach feature is used by the terminal to notify the network whether the terminal is reachable again. The network will modify the status information of the terminal. The detach feature is used by the terminal to notify the network whether the terminal is temporarily not reachable.

- *Terminal Paging* enables the determination of the current location of a user or of a mobile terminal.
- *Radio paging* enables one way personal selective calling with alert. This feature enables a user to send a message, either voice, tone, or alphanumeric, to a selected pager terminal or a group of terminals.
- *Emergency Calls in Wireless* allows emergency calls to have priority over all other calls to ensure service.
- *Terminal Equipment Validation* should be considered as a part of the mobility processes of Location Management (i.e., terminal location registration), Call Origination, Call Delivery and at other times as initiated by the network or terminal.

## B.1.5 Other services and service features

- *Multimedia* allows a subscriber to receive or send an integrated communication consisting of mixtures of voice, data, image, and video information. A key capability will be the ability to synchronise and control delivery of information from disparate sources (e.g. voice and data). This will include controlling delivery from multiple sources to a single recipient and from a single source to multiple recipients.
- *Message store and forward* enables a user to send a message to be distributed to one or several destination users. Different types of messages may be supported, e.g. voice, data, fax, and different methods of delivery and/or times of delivery may be specified, e.g. only to pre-subscribed mail-box holders, or direct to any access.
- *Hot Line* allows a user to place calls without providing, in the call request, the called party information required by the network to route the call. This routing information is stored in the network by prior subscription.
- *Terminating Key Code Screening Protection* enables a subscriber to protect his line and screen incoming calls by means of a user defined key, i.e., pin code. The callers are required to enter this key.

The following service features are envisaged for supporting the above services and also other services:

- *Calling Name Delivery* gives to the network operator the capability to display/announce the name of the calling party to the calling name delivery user (the called party) prior to answer, thus allowing this user to screen or distinctively answer the call.

- *Services on-demand* enables a user to request new services while initi-ating or involved in a call, e.g., multi-way calling at a pay phone. This includes the capability to invoke new services for the duration of the call.

- *Message Waiting Indication* enables a user to be informed that messages for his attention are waiting.

- *Feature use charging* enables the service provider to apply a certain charge to the use of any specified feature.

- *Resource Allocation* enables the allocation, in advance and for a certain period of time, of pooled resources (e.g., conference bridges) required for a service.

- *Special Facility Selection* enables a call to be routed via a special facility (e.g. a virtual leased line) under the determination of service control.

- *Service Indication* enables the called party to receive an indication concerning the presented call (for instance, an application to the Freephone service would be the indication that the charge is to be supported by the called party; an application to the Call Forwarding service would be the forwarding number).

- *Service negotiation* enables the parties involved in a call to negotiate the bearer services, teleservices and supplementary services to be provided for the call, depending on the services subscribed by the parties, the terminal and network capabilities, etc. This negotiation may take place both during call set up and during the active call phase.

- *User Service Interaction* enables a user to interact with a service, and thus to send or receive information to or from a service in association with a call.

- *Delivery of Complementary Information* enables a calling user to supply to the network complementary information (e.g., an account number and a password) associated with the call set-up information.

- *Call Connection Elapsed Time Limitation* allows a calling party to make calls and communicate with one or more parties within a duration time predefined on a subscription basis.

- *Concurrent Features Activation with Bi-Control* enables a call to be influ-enced by some features at two different points concurrently, such as originating side and terminating side.

- *Customised Call Routing with customers private networks* permits the public network to access private networks customers systems for call processing and routing information. The accessed private network system determines for each incoming call the appropriate destination, which could be a local, national, or international telephone number.

- *Data communication between different protocol terminals* enables a mobile terminal to handle data communication between different protocol terminals in the intra-network or inter-network environment. This service feature is performed by corresponding protocol conversion units within an IP upon requesting the data communication from the subscriber.
- *Charge determination* enables the calculation of charges related to a call.
- *Enhanced Call Disposition* provides a means to cut down the call as soon as the (calling) card usage is exceeded.
- *B-ISDN Multiple connections point to point* enables a user to make a call between two points involving multiple connections, e.g., voice, audio, video, and/or data. This service feature may make use of B-ISDN Point to Point Connection service feature and needs other service features such as multi connection, etc.
- *B-ISDN Multi-casting* enables the network to set up multiple connections among multiple parties where the connections are point to multi-point unidirectional. This service feature may make use of B-ISDN Point to Multipoint Connection service feature and/or B-ISDN Multipoint to Point Connection service feature and needs other service features such as leaf control, etc.
- *B-ISDN Conferencing* enables the network to set up multiple connections among multiple parties where the connections are multi-point to multi-point. This service feature may make use of B-ISDN Multipoint to Multipoint Connection service feature and other service features such as third party control, etc.

## B.2 CS-2 service creation and service management services

In addition to telecommunications services, CS-2 addresses initially service creation services and service management services. A complete specification of these services is expected to be completed within CS-3. Therefore we indicate only briefly the scope of these service categories.

'Service creation services' encompass service specification services, service development services, service verification services, and service creation service management services.

'Service management services' encompass the activities performed by network operators and service providers during the deployment, provisioning and utilization phases of an IN service. After service deployment, service customization services, service control services and service monitoring services will be used for service provisioning and utilization. Generally the services are described as follows:

- Service deployment is the introduction of a service into the IN-structured network in a subscriber independent way;
- Service provisioning is the initial installation and deployment of necessary resources and data in appropriate network elements to provide a service subscription to a specific subscriber along various service customisation parameters values and the initial activation;
- Management during the service utilization contains service monitoring, service maintenance, service traffic management, audit administration, and billing activities.

# Abbreviations

| | |
|---|---|
| ACTS | advanced communications technologies and services |
| AIN | advanced intelligent network |
| AINAP | advanced intelligent network application protocol |
| ANSI | American National Standards Institute |
| API | application programming interface |
| ASN.1 | abstract syntax notation no. 1 |
| ATM | asynchronous transfer mode |
| BCAF | bearer control agent function |
| BCF | bearer control function |
| BCP | basic call process |
| BCSM | basic call state model |
| BCUP | Basic Call Unrelated Process |
| B-ISDN | broadband ISDN |
| CAMEL | customized applications for mobile network enhanced logic |
| CCAF | call control agent function |
| CCITT | Comitté Consultatif Internationale de Telegraphique et Telephonique |
| CCF | connection control function |
| CCS7 | common channel signaling system no. 7 |
| CMIPS | common management information protocol service |
| CPE | customer premises equipment |
| CT2 | cordless telephony (second generation) |
| CTM | cordless terminal mobility |
| CID | call instance data |
| CS-1 | capability set 1 |
| CS-2 | capability set 2 |

| | |
|---|---|
| CS-3 | capability set 3 |
| CPE | customer premises equipment |
| CSM | connection session manager |
| DAP | directory application protocol |
| DCS1800 | digital communications systems at 1800 MHz |
| DECT | digital European cordless telecommunication |
| DFP | distributed functional plane |
| DP | detection point |
| DPE | distributed processing environment |
| DSL | distributed service logic |
| EDP | event detection point |
| ETSI | European Telecommunications Standardization Institute |
| EURESCOM | European Institute for Research and Strategic Studies in Telecommunications |
| E-OSF | operations system function for (network) element management |
| f | (TMN) f-type reference point |
| FE | functional entity |
| FEA | functional entity action |
| FPLMTS | future public land mobile telecommunications system |
| GSEP | generic session end point |
| GFP | global functional plane |
| GSL | global service logic |
| GSM | global system for mobile communications |
| HLSIB | high-level SIB |
| IF | information flow |
| IN | intelligent network |
| INA | information networking architecture |
| INAP | intelligent network application protocol |
| INCM | intelligent network conceptual model |
| IP | intelligent peripheral |
| ISDN | integrated services digital network |
| ISO | International Standardization Organization |
| ISUP | ISDN user part |
| ITU | International Telecommunications Union |
| IWF | interworking function |
| LEX | local exchange |
| MACF | multiple association control function |
| MAP | mobile application part |
| MO | managed object |

| | |
|---|---|
| MTP | message transfer part |
| MVIF | Multi-Vendor Interaction Forum |
| NAP | network access point |
| NCCE | native computing and communications environment |
| NCSF | non-call service function |
| NEF | network element function |
| N-OSF | operations system function for network management |
| O_BCSM | originating basic call state model |
| ODP | open distributed processing |
| OAMP | operations, administrations and maintenance part |
| OSF | operations system function |
| OSI | open systems interconnection |
| PE | physical entity |
| PIC | point in call |
| POI | point of initiation |
| POR | point of return |
| POS | point of synchronization |
| POTS | plain old telephone service |
| PP | physical plane |
| PPrf | personal profile |
| PSTN | public switched telephone network |
| QAF | Q-Adaptor Function |
| q3 | (TMN) q3-type reference point |
| RAE | Research Programme for Advanced Communications in Europe |
| RCAF | radio control access function |
| RLF | radio link function |
| RM-ODP | reference model for open distributed processing |
| ROSE | remote operations service element |
| SACF | single association control function |
| SCCP | signaling connection control part |
| SCEF | service creation environment function |
| SCEP | service creation environment point |
| SCF | service control function |
| SCP | service control point |
| SCSO | service core support object |
| SCUAF | service control user agent function |
| SDF | service data function |
| SDL | service description language |

| | |
|---|---|
| SDP | service data point |
| SF | service feature (service factory in TINA section) |
| SIB | service-independent building block |
| SLP | service logic program |
| SMAF | service management agent function |
| SMAP | service management agent point |
| SMF | service management function |
| SMP | service management point |
| SMS | service management system |
| SRF | specialized resource function |
| SSCP | service switching and control point |
| SSD | service support data |
| SSF | service switching function |
| SSM | service session manager |
| SSP | service switching point |
| SSCO | service substance support object |
| STP | signaling transfer point |
| SubAgt | subscription agent |
| S-OSF | operations system function for service management |
| TE-A | terminal equipment agent |
| TEX | transit exchange |
| T_BCSM | terminating basic call state model |
| TCAP | transaction capabilities application part |
| TSA | terminal service adapter |
| TDP | trigger detection point |
| TINA | telecommunication information networking architecture |
| TINA-C | TINA consortium |
| TMN | telecommunication management network |
| TUP | telephone user part |
| UA | user agent |
| UAP | user application |
| UCxt | usage context |
| UMTS | universal mobile telecommunications system |
| USM | user session manager |
| UPT | universal personal telecommunications |
| VPN | virtual private network |
| WSF | workstation function |
| x | (TMN) x-type reference point |

# Glossary

**Adjunct**   A physical element in the INCM physical plane that is functionally equivalent to an SCP but is directly connected to an SSP.

**Advanced intelligent network (AIN)**   An evolvable IN architecture developed by Bellcore and the MVIF as the successor to the IN-1+ architecture in the US. AIN can be regarded as a counterpart to the ITU IN recommendations in the US.

**Basic call process (BCP)**   A specific SIB that is responsible for providing basic call connectivity between parties in the network and allows the passing of call control temporarily to an IN service. Thus, the BCP includes points of initiation (POI) to GSL and points of return (POR) from GSL to the BCP. The BCP is closely related to the BCSM.

**Basic call state model (BCSM)**   A high-level finite-state machine description of switch (i.e. CCF) activities required to establish and maintain communication paths. The BCSM is defined by the sets of states that are used to illustrate the different states that a call can go through, from call origination (i.e. the calling user picks up the phone) to call termination (i.e. the user hangs up).

**Basic Unrelated Process (BCUP)**   A specific SIB within the CS-2 GFP which provides the non call associated capabilities. The BCUP is an independent process. The concept is similar to that of the BCP.

**Bearer control**   The set of functions used to direct the lower layer means of transmission.

**Bearer network**   Provides for basic transmission of information. In the IN context the bearer network is modeled by interconnected CCFs that cooperate with each other to provide network call processing functions.

**Call control**   The set of functions used to process a call, including establishment, supervision, maintenance, release of connections, and service feature provision.

**Call control agent function (CCAF)**   A functional entity providing the user with access function to the network in the INCM DFP. It is the interface between the user and the network connection control functions.

**Call party**   The entity that receives a call from a service.

**Calling party**   The entity that originates a call to a service.

**Call instance data (CID)**   An identifier that defines subscriber specific details for SIBs in the GFP.

**Capability set (CS)**   Defines a specific standardized stage of IN evolution in terms of the services to be supported and the functional architecture supporting these services. CS-1 (released in 1992) is the first set of international IN standards.

**Common channel signaling system no. 7 (CCS7)**   A packet-switched 'out-of-band' signaling network for the transport of circuit-related information between exchanges for set-up and release of calls.

**Connection control function (CCF)**   A functional entity in the INCM DFP providing the call processing and control in a bearer (telecommunication) network for basic telephony services and also advanced, switch-based services. The CCF runs the BCSM (on both the originating and terminating ends of a call) in order to allow IN service invocation.

**Core feature**   A service feature that is fundamental for a service, i.e. in the absence of the feature the service is not a viable service offering.

**Core INAP**   A reviewed and refined version of the ITU INAP, defined by the ETSI for the European context.

**Cordless terminal mobility (CTM)**   An IN service currently under definition in the ETSI, providing cordless telephones (i.e. CT2 or DECT) to be connected to public IN-structured networks. In particular, internetworking between private and public networks will be supported in order to support user roaming.

**Detection point (DP)**   A point in basic call processing, i.e. in the BCSM, at which an event may be reported to an SCP and transfer of processing control to IN service logic can occur. Two types of DPs are defined: event DPs and trigger DPs.

**Distributed functional plane (DFP)**   The third plane of the INCM, this models the distributed realization of IN services by means of a set of IN functional entities that interact via information flows for the realization of SIBs.

**Distributed processing environment (DPE)**   A computing platform that hides from applications, i.e. computational objects representing service components, the underlying technologies and distribution concerns, thus supporting the construction of portable and interoperable code.

**Distributed service logic (DSL)**   The service logic in the DFP used to realize SIBs, taking into account distribution aspects of the IN functional architecture. Hence, it consists of service trigger information, service logic program, service data, and optionally specialized resource data.

**Event**   A specific input to and/or output from a given state in a finite-state machine model, i.e. a point in call (PIC), that causes a transition from one state to another.

**Event detection point (EDP)**   A detection point in basic call processing that is dynamically armed and used to reporting the occurrence of a specific event to a running service logic program.

**Freephone (FPH)**   Also known as 'Green Number' service or 'Toll Free' service, this is a classical IN service in which the call charges are allocated to the called party, representing the service subscriber (usually a business company).

**Functional entity (FE)**   An IN network element involved in IN service provision and located in the DFP of the INCM.

**Functional entity action (FEA)**   An action performed by a functional entity as a result of a specific stimulus while the functional entity is in a specific state.

**Future public land mobile telecommunications system (FPLMTS)**   An international standard for the third-generation mobile telecommunications systems, integrating current cordless, cellular, paging systems, as well as UPT and mobile satellite systems, aiming for broadband transmission capabilities. FPLMTS is also referred to as IMT-2000.

**Global functional plane (GFP)**   The second plane of the INCM, supporting the service designer view, which models the (IN) network as a single programmable entity, hiding the distributed nature of the network.

**Global service logic (GSL)**   The logic in the GFP to achieve IN service features, describing how SIBs are chained together and the interactions between the BCP SIB and these SIB chains.

**Global system for mobile communications (GSM)**   A European standard for a cellular mobile telecommunications network, enabling voice and data transmission.

**High-level SIB (HLSIB)**   Defined in CS-2, this is an SIB that is composed of other HLSIBs and normal SIBs. HLSIBs, like normal SIBs, are reusable parts of a service feature.

**Home network**   The IN where the service subscriber is administered and where the service logic and data, particularly the customer data (customer profile), resides.

**Information flow (IF)**   A client–server relationship between two functional entities in the DFP required in cases where different functional entities have to communicate to perform specific SIB functionality.

**Intelligent network (IN)**   An architectural concept for the creation and provision of advanced telecommunication services. Its architecture can be applied to various types of telecommunication networks, including PSTN, N-ISDN and B-ISDN.

**Intelligent network application protocol (INAP)**   Defines the interface protocol for the communication between IN physical network elements (e.g. between SSP and SCP), based on CCS7. INAP is defined in ITU-T. A European version, known as 'core INAP', is defined by the ETSI.

**Information networking architecture (INA)**   An integrated architecture for future telecommunications and operations services driven by IN, TMN, and ODP concepts, developed by Bellcore as the successor to its AIN architecture.

**Intelligent network conceptual model (INCM)**   A modeling tool for the creation of IN architectures that consists of four planes addressing service design aspects, global and distributed provision functionality, and physical aspects of INs.

**Intelligent peripheral (IP)**   Adds additional capabilities to a switch, such as digit recognition and speech synthesis and is an implementation of an SRF.

**IN service**   A telecommunication service offered to end users by a service provider in which the service logic is executed to provide the service. A service constitutes a stand-alone commercial offer, characterized by one or more *core* service features, that can be optionally enhanced by other service features.

**Interworking function (IWF)**   A CS-2 functional entity that provides the specific capabilities for IN interworking.

**Interworking service**   A service which is provided across interconnected IN platforms.

**Integrated services digital network (ISDN)**   A digital transmission network enabling integrated voice and data transmission.

**Local exchange**   Also known as 'local central office', this is an exchange in which subscriber lines terminate. It is the lowest level of the switching hierarchy.

**Network access point (NAP)**   A physical entity containing the CCAF and CCF, which provides network access to users.

**Non-call service function (NCSF)**   A CS-2 functional entity that provides a set of functions required for access and interaction between the user and the SCF for non-call associated services, such as mobility service features. This functional entity may be renamed into Call Unrelated Service Function, CUSF.

**Open distributed processing (ODP)** An international standard that goes beyond the OSI reference model by defining a framework for the definition of open distributed systems and distributed applications based on the object-oriented paradigm.

**Open network provision (ONP)** A legal framework for the establishment of a uniform telecommunications infrastructure in the European Community based on an open market of telecommunications services and equipment. The basic idea is to establish open interfaces and equal use conditions for the public network infrastructure.

**Operations system function (OSF)** Provides the TMN functions for processing, storage, and retrieval of management information, forming the core part of a the target TMN. OSFs can be allocated to four hierarchical layers, namely business management, service management (S-OSF), network management (N-OSF), and network element management (E-OSF).

**Originating network** Provides the calling party with the access to the IN and to the IN services. In order to provide this service access, some specific service logic and data is required in the originating network.

**Plain old telephone service (POTS)** The basic telephone service, requiring nothing more than basic call handling.

**Point in call (PIC)** A state in the BCSM providing an external view of a call-processing state or event to IN service logic.

**Point of initiation (POI)** A point in basic call process SIB at which a service logic program begins, i.e. it marks the point at which control is transferred from the CCF to the SCF. The POI is associated with a DP in the BCSM and provides new call state information and parameters for use by subsequent processing.

**Point of return (POR)** A point in basic call process SIB at which an IN service logic program returns control from the SCF to the CCF. The POR is associated with a DP in the BCSM and provides new call state information and parameters for use by subsequent processing.

**Point of synchronization (POS)** Defined in CS-2, this is a point in service processing at which synchronization between two IN service processes takes place.

**Premium rate (PRM)** A service defined within IN CS-1, enabling a service subscriber with a premium rate number to obtain extra charges for the provision of special services/information to calling users. Users calling a premium rate number will be charged at a special rate both for the call and for the information/service obtained by the call. The network operator collects the revenue and shares it with the service/information provider.

**Public switched telephone network (PSTN)** A telecommunications network set up to perform telephony services for the public subscribers.

**Service access code** The first part of an IN service number. It is used for recognition of IN service calls within the exchanges, i.e. CCF/SSF.

**Service control** The direction of the functions or processes used to provide a specific telecommunications service.

**Service control function (SCF)** An IN functional entity in the DFP which is the heart of the IN and controls resources in a switch and/or intelligent peripheral. The SCF contains the service logic programs that are used to provide IN services.

**Service control point (SCP)** An implementation of a SCF in the physical plane of the INCM.

**Service control user access function (SCUAF)** A CS-2 functional entity that provides user access to the non-call-associated service capabilities of the NCSF, required for mobility service features.

**Service creation** The complete process of IN service creation, including design, specification, development, and verification.

**Service creation environment function (SCEF)** An IN functional entity providing software engineering tools that support the IN service creation process, resulting in output of both service logic and service data (template).

**Service creation environment point (SCEP)** A physical entity that implements the SCEF.

**Service data** Customer and/or network information required for the proper functioning of a service.

**Service data function (SDF)** An IN functional entity in the DFP that contains the service data (customer- and network-related data) and provides standardized real-time access for SCFs to service data. Sometimes referred to as 'specialized data function'.

**Service data point (SDP)** An implementation of an SDF in the physical plane of the INCM. Sometimes referred to as 'specialized data point'.

**Service data template (SDT)** A data template related to a specific service logic program, required for the SDF.

**Service feature (SF)** Reflects a specific aspect of the functionality of an IN service. A service feature is either a core part of a telecommunication service or an optional part offered as an enhancement to a telecommunication service.

**Service-independent building block (SIB)**   A unit of network functionality which can be reused by service logic for customized service control. A SIB is the building block used to create global service logic.

**Service logic (SL)**   A sequence of processes/functions used to provide a specific IN service.

**Service logic program (SLP)**   A software program containing the service logic that runs in an SCP.

**Service management agent function (SMAF)**   An IN functional entity in the DFP providing the man–machine interface to the SMF.

**Service management agent point (SMAP)**   A physical entity that implements the service management agent function.

**Service management function (SMF)**   A functional entity in the DFP of the INCM which provides capabilities for operations that support IN services in the IN architecture defined in CS-1.

**Service management point (SMP)**   A physical entity that implements the service management function.

**Service management system (SMS)**   An operations system for IN service management.

**Service number**   Comprises usually an IN service access code plus a subscriber-specific number. The former is used to determine a corresponding service logic program (i.e. SCP), the latter for determination of the corresponding subscriber data.

**Service plane**   The uppermost plane of the INCM describing services and service features from a user´s perspective.

**Service process**   Introduced within CS-2 to allow for parallel execution of GSL, i.e. several threads of SIB chains. GSL is formed by several SIB chains, namely service processes, that execute different activities related to the same service.

**Service provider**   An organization that commercially offers and manages IN services.

**Service subscriber**   An entity that makes use of IN services offered by service providers.

**Service support data (SSP)**   An identifier that defines data parameters of specific service features descriptions for SIBs in the GFP.

**Service switching function (SSF)**   An IN functional entity in the DFP that is additional functionality for controlling switch resources and provides a well-defined, service-independent interface between the CCF and the SCF. The SSF contains detection capabilities, e.g. service trigger information, which detect requests for IN services.

**Service switching and control point (SSCP)**   A physical entity in the physical plane of the INCM that contains the CCF/SSF, SCF, and SDF. Thus, it is a combined SSP and SCP.

**Service switching point (SSP)**   A physical entity in the physical plane of the INCM implementing an SSF.

**Service trigger information (STI)**   Stimulus information that initiates an action in the CCF/SSF. It may be distinguished from a 'trigger detection point' initiating a service logic program and an 'event detection point' reporting the occurrence of a specific event to a running service logic program.

**Service user**   An entity external to the network that uses its services.

**Session**   A term used to replace the traditional notion of a 'call' within TINA. A session is the purpose of a service which is achieved by performing a collection of activities during a specific period of time.

**Signaling transfer point (STP)**   A very high-capacity, very reliable packet switch in the CCS7 network that transport signaling messages between the network nodes, e.g. SSPs and SCPs.

**Specialized resource function (SRF)**   An IN functional entity in the DFP that provides additional functions to a switch for controlling (intelligent) peripheral resources, e.g. for collecting information from call parties or playing announcements to call parties.

**Supplementary service**   A basic service enhanced by additional service features or capabilities.

**Transaction capabilities application part (TCAP)**   A specific user part in the CCS7 network that provides remote operation capabilities to higher signaling application protocols, such as INAP. TCAP uses the CCS7 transport services provided by MTP and SCCP.

**Telecommunication management network (TMN)**   A synonym for the ITU M.*xxxx* series recommendations. Alternatively, it denotes an existing management network belonging to one legal authority.

**Telecommunication information networking architecture (TINA)**   An architecture for future telecommunications and operations services driven by IN, TMN, and ODP. TINA is being developed by an international consortium of network operators and equipment manufacturers exploring the future evolution of telecommunication environments.

**Televoting (VOT)**   A service defined within IN CS-1 enabling the subscriber to perform phone voting by allocating one or more temporary numbers. The service may provide facilities to play announcements acknowledging the call, to count the calls and to relate the total number of calls registered.

**Terminating network**   A telecommunications network providing access to the called party.

**Trigger**   A stimulus that initiates an action.

**Trigger detection point (TDP)**   A detection point in basic call processing that is statically armed and used for initiating a service logic program.

**Type A services**   Defined within IN CS-1 as representing two-party, single bearer and single point of control services that can only be activated during the set-up or tear-down phase of a call. Services without these limitations are referred to as type B services.

**Universal access number (UAN)**   A service defined within IN CS-1 enabling the terminating lines of a subscriber at different locations to be reached via a unique number. Based on the area from which the incoming call originates, the subscriber may specify a number to which the call is routed (e.g. the nearest office).

**Universal personal telecommunications (UPT)**   A service defined within IN CS-1 enabling access to a user-defined set of telecommunication services supporting personal mobility and service personalization. Service subscription is based upon an unique, personal, network-transparent UPT number.

**Universal mobile telecommunications system (UMTS)**   A European standard for the third-generation mobile telecommunications system, integrating current cordless, cellular, paging systems, as well as UPT and mobile satellite systems, with the goal of broadband transmission capabilities. UMTS is closely related to ITU FPLMTS (IMT-2000).

**User agent (UA)**   A computational object of the TINA service architecture. The UA is a user in the service provider domain and acts as a single access point to control and manage user/service sessions.

**Virtual private network (VPN)**   A service defined within IN CS-1 providing private network capabilities by using public network resources. The subscriber's lines, connected to different network switches, constitute a virtual private network including capabilities such as private numbering plan and call transfer.

**Workstation function (WSF)**   A functional building block that provides access to a TMN OSF and thus provides the user interface to an OSF.

# References

Ambrosch WD *et al.*: *The Intelligent Network, a Joint Study by Bell Atlantic, IBM and Siemens*. Springer, Heidelberg, 1989.

Bagely M *et al.*: The information services supermarket – a trial TINA-C design, in *Proceedings of the 5th TINA Workshop*, Melbourne, February 1995, pp 765–777.

Barr W *et al.*: The TINA initiative. *IEEE Communications Magazine*, 1993, 31(3), pp 70–77.

Basu K: Open network architecture and information services, in *Computer Networks and ISDN Systems*, Vol 20. North Holland, 1990, pp 101–107.

Beires N, T Magedanz, M Kockelmans: An evolutionary approach for the TMN management of IN services, in *Lecture Notes on Computer Science 851 – Toward a Pan-European Telecommunication Service Infrastructure – IS&N'94*, Kugler *et al.* (eds). Springer, 1994, pp 285–294.

Bellcore Special Report: Plan for the second generation of intelligent network, Issue 1, July 1986.

Bellcore Special Report: IN/1+, network baseline architecture, SR-NPL-001052, May 1988.

Bellcore Special Report: Advanced intelligent network release 1 proposal, SR-NPL-001509, Issue 1, November 1989.

Bellcore Special Report: Advanced intelligent network release 1 network and operations plan, SR-NPL-001623, Issue 1, June 1990.

Bellcore Technical Report: AIN 0.1 switching system generic requirements, TR-NWT-001284, August 1992a.

Bellcore Technical Report: AIN 0.1 switching system–service control point application protocol interface generic requirements, TR-NWT-001285, August 1992b.

Bellcore Special Report: Cycle 1 initial specifications for information networking architecture (INA), SR-NWT-002268, Issue 1, June 1992c.

Bellcore Technical Report: Advanced intelligent network (AIN) service control point (SCP) generic requirements, TR-NWT-001280, August 1993a.

Bellcore Special Report: Advanced intelligent network (AIN) operations system (OS)–service control point (SCP) interface generic requirements, GR-1286-CORE, Issue 1, September 1993b.

Bellcore Special Report: Cycle 1 specifications for information networking architecture (INA), SR-NWT-002268, Issue 2, April 1993c.

Bellcore Special Report: AIN 0.2 switching system requirements, GR-1298-CORE, Issue 1, November 1993d.

Bellcore Special Report: AIN 0.2 switching system–service control point (SCP)adjunct application protocol interface generic requirements, GR-1299-CORE, Issue 1, November 1993e.

Bellcore Special Report: AIN 0.2 switch–intelligent peripheral generic requirements, GR-1229-CORE, Issue 1, November 1993f.

Bellcore Special Report: AIN service control point (SCP) generic requirements, GR-1280-CORE, April 1994 (replaces Bellcore 1993a)

Bellcore Special Report: AIN switching system requirements, GR-1298-CORE, Issue 2, December 1994a.

Bellcore Special Report: AIN switching system–service control point (SCP)/ adjunct application protocol interface generic requirements, GR-1299-CORE, Issue 2, December 1994b.

Berndt H, *et al.*: Service and management architecture in TINA-C, in Proceedings *of the 5th TINA Workshop*, Melbourne, February 1995, pp 81–96.

Brown DK: Practical issues involved in architectural evolution from IN to TINA, in *International Conference on Intelligent Networks (ICIN)*, Bordeaux, October 1994, pp 270–274.

Bretecher Y, B Vilian: The intelligent network in a broadband context, in *XV International Switching Symposium*, Berlin, April 1995.

Cancer E, *et al.*: IN rollout in Europe. *IEEE Communications Magazine*, March 1993, 31(3), 38–47.

Carl D, *et al.*: Intelligent network implementation examples, in *Proceedings of the IEEE IN Workshop*, Heidelberg, March 1994, paper 112.1.

Cavalho P, *et al.*: Service creation environment for IN CS1, in *Proceedings of the 3rd IEEE International Workshop on Intelligent Networks*, Heidelberg, May 1994, paper 212.3.

Chabernaud C, B Vilain: Telecommunication services and distributed applications. *IEEE Network Magazine*, November 1990, 10–13.

Covaci S, T Magedanz, N Beires, I Kruzela: Toward pan-European IN services – IN interworking and IN management issues, in *XV International Switching Symposium (ISS)*, Berlin, April 23–28, 1995, VDE, pp 112–116.

Dang JC, *et al.*: IN as a platform for UPT: constraints and requirements, in *2nd International Conference on Intelligence in Networks*, Bordeaux, March 1992, pp 79–82.

Drignath R, *et al.*: Alcatel's intelligent network product line, in *Proceedings of the IEEE IN Workshop*, Heidelberg, March 1994, paper 112.2.

EC: Council Directive 90/381/EEC of 28 June 1990 on the establishment of the internal market for telecommunications services through the implementation of open network provision, Brussels, 28 June 1990.

Eckardt T, T Magedanz, R Popescu-Zeletin: Application of X.500 and X.700 standards for supporting personal communications in distributed computing environments, in *5th IEEE Computer Society Workshop on Future Trends of Distributed Computing Systems (FTDCS)*, Cheju Island, South Korea, August 1995, pp 232–241.

Garrahan JJ, *et al.*: Intelligent network overview. *IEEE Communications Magazine*, March 1993, 31(3), 30–37.

Griffeth ND, H. Velthuijsen: The negotiating agents approach to runtime feature interaction resolution, in *Feature Interactions in Telecommunications Systems*, LG Bouma and H Velthuijsen (eds). IOS Press, Amsterdam, 1994.

Händel R, *et al.*: *ATM Networks – Concepts, Protocols, Applications*, 2nd edn, Addison-Wesley, 1994.

Hecker HPJ, *et al.*: The application of the IN-concept to provide mobility in underlying networks, in *2nd International Conference on Intelligence in Networks (ICIN)*, Bordeaux, March, 1992, pp 95–100.

Hetz H: 1993–1996 study period scope of work for intelligent networks, ITU Study Group XI, Geneva, September 1992.

Hu YC: IN CS-2 enhancements to the global functional plane, in *Proceedings of the 4th IEEE International Workshop on Intelligent Networks*, Ottawa, May 1995, paper 223.3.

Iida I, *et al.*: DUET: Agent-based personal communications network, in *XV International Switching Symposium*, Berlin, April 1995, VDE, pp 119–123.

Niebert N, E Geulen: Personal communications – what is beyond radio, in *Lecture Notes on Computer Science 851 – Toward a Pan-European Telecommunication Service Infrastructure – IS&N'94*, Kugler *et al.* (eds). Springer, 1994, pp 247–258.

IEEE: Intelligent networks: toward implementation in fixed and mobile networks. *IEEE Communications Magazine*, February 1992, 1(2).

IEEE Special issue: PCS: the second generation. *IEEE Communications Magazine*, December 1992, 30(12).

IEEE Feature topic: Managing feature interactions in telecommunications systems. *IEEE Communications Magazine*, August 1993a, 31(8).

IEEE Special issue: Marching toward the global intelligent network. *IEEE Communications Magazine*, March 1993b, 31(3).

IEEE Feature topic: Enter the world of TMN. *IEEE Communications Magazine*, March 1995, 33(3).

*Net Telecommunications Information Networking Architecture (TINA) Workshop*, Lake Mohonk, NY, June 1990.

Kelly E, *et al.*: TINA-C DPE architecture and tools, in *Proceedings of the 5th TINA Workshop*, Melbourne, February 1995, pp 39–54.

Kitson B: CORBA and TINA: the architectural relationship, in *Proceedings of the 5th TINA Workshop*, Melbourne, February 1995, pp 371–386.

Kitson B, P Leydekkers *et al.*: TINA object definition language (TINA-ODL) manual, version 1.3, TINA-C stream deliverable, 20 June 1995.

Kockelmans M, E de Jong: Overview of IN and TMN harmonization, *IEEE Communications Magazine*, March 1995, 33(3), 62–67.

Lauer G, *et al.*: Broadband intelligent network architecture, in *Intelligent Network Workshop (IN'95)*, Ottawa, May 1995, paper 212.3.

Leconte A, *et al.*: IN value added – the role of the IP/SN in IN and multimedia applications, in *XV International Switching Symposium*, Berlin, April 1995, VDE, pp 206–210.

Magedanz T: IN and TMN providing the basis for future information networking architectures. *Computer and Communications*, May 1993, 16(5), 267–276.

Magedanz T: An integrated management model for intelligent networks, PhD Thesis, GMD-Bericht 217. R. Oldenbourg, Germany, January 1994.

Magedanz T: On the integration of IN and TMN – modeling IN-based service control capabilities as part of TMN-based service management, in *Integrated Network Management*, Vol IV, AS Sethi *et al.* (eds). Chapman & Hall, London, 1995, pp 386–397.

Magedanz T, R Popescu-Zeletin: Applying open network provision to ISDN and IN. *Computer Networks and ISDN Systems*, March 1992, 24, 1–14.

Magedanz T, *et al.*: Managing intelligent networks the TMN way: IN service versus network management, in *RACE International Conference on Intelligence in Broadband Service and Networks (IS&N)*. Paris, November 1993, paper IV/1.

Magedanz T, *et al.*: Intelligent agents: an emerging technology for next generation telecommunications, in *IEEE INFOCOM'96*, San Francisco, March 1996.

Martin M: IN architecture: an opportunity for providing new services in GSM networks, in *International Conference on Intelligent Networks (ICIN)*, Bordeaux, October 1994, pp˙311–331.

Minzer S, *et al.*: Evolutionary trends in call control, in *XV International Switching Symposium (ISS)*, Berlin, April 1995, VDE, pp 300–304.

Mitra N, SD Usiskin: Relationship of the signaling system no. 7 protocol architecture to the OSI reference model. *IEEE Network Magazine*, January 1991, 26–37.

Modaressi A, R Skoog: Signalling system no. 7: a tutorial. *IEEE Communications Magazine*, July 1990, 19–35.

Mouly M, MB Pautet: *The GSM System for Mobile Communications*. M Mouly & MB Pautet Publishers, 1992.

Natarajan N, G Slawsky: A framework architecture for information networks. *IEEE Communications Magazine*, February 1992, 1(2), 102–109.

Natarajan N: INAsoft DPE: a platform for distributed telecommunications applications, in *Proceedings of the 5th TINA Workshop*, Melbourne, February 1995, pp 67–80.

van Nielen M: The impact of mobility on intelligent network functions, in *International Zurich Seminar on Intelligent Networks and their Applications*, Zürich, March 1992, pp 29–40.

Nilsson G, *et al.*: An overview of the telecommunications information networking architecture, in *Proceedings of the 5th Telecommunications Information Networking Architecture (TINA) Workshop*, Melbourne, February 1995, pp 1–12.

*Proceedings of the 3rd IEEE International Workshop on Intelligent Networks*, Heidelberg, May 1994.

*Proceedings of the 4th IEEE International Workshop on Intelligent Networks*, Ottawa, May 1995.

*Proceedings of the 5th IEEE International Workshop on Intelligent Networks*, Melbourne, April 1996.

*Proceedings of the 2nd International Conference on Intelligence in Networks*, Bordeaux, March, 1992.

*Proceedings of the 3rd International Conference on Intelligence in Networks*, Bordeaux, October 1994.

*Proceedings of the Third International Workshop on Feature Interactions in Telecommunications Software Systems*, Kyoto, October 1995.

Rapeli J: UMTS targets, system concept and standardization in a global perspective. *IEEE Personal Communications*, February 1995, 2(1), 18–27.

Reinhardt A: The network with smarts. *BYTE Magazine*, October 1994, 51–64.

Rizzo M, IA Utting: An agent-based model for the provision of advanced telecommunication services, in *Proceedings of the 5th Telecommunications Information Networking Architecture (TINA) Workshop*, Melbourne, February 1995, pp 205-218.

Rubin H, N Natarajan: A distributed software architecture for telecommunication networks. *IEEE Network*, January/February 1994, 8(1), 8–17.

Rumbaugh J, *et al.*: *Object-Oriented Modeling and Design*. Prentice Hall, 1991.

Russo PA, *et al.*: IN rollout in the United States. *IEEE Communications Magazine*, March 1993, 31(3), 56–63.

Suzuki S: IN rollout in Japan. *IEEE Communications Magazine*, March 1993, 31(3), 48–55.

*2rd Telecommunications Information Networking Architecture (TINA) Workshop*, Chantilly, March 1991.

*3rd Telecommunications Information Networking Architecture (TINA) Workshop*, Narita, Japan, January 1992.

*4th Telecommunications Information Networking Architecture (TINA) Workshop*, L'Aquila, Italy, September 1993.

*5th Telecommunications Information Networking Architecture (TINA) Workshop*, Melbourne, February 1995.

Turner DG: Status and future directions for global intelligent network standards, in *Proceedings of the 4th IEEE IN Workshop*, Ottawa, May 1995, paper 223.1.

Tuttlebee W: Cordless personal communications. *IEEE Communications Magazine*, December 1992, 30(12), 42–54.

Yagi H, *et al.*: TINA-C service components, in *Proceedings of the 5th Telecommunications Information Networking Architecture (TINA) Workshop*, Melbourne, February 1995, pp 113–124.

# Standards and recommendations

CCIR Recommendation 816: Framework for services supported on future public land mobile telecommunication system (FPLMTS), question 39/8, 1992.

CCIR Recommendation 817: future public land mobile telecommunication system (FPLMTS) – network architectures, question 39/8, 1992

ETSI DTR SMG-50301: Framework of network architecture, interworking and integration for the universal mobile telecommunications system (UMTS), version 1.5.1, June 1995.

ETSI Draft DE NA-10039: CTM service description – phase 1, November 1995.

ETSI DTR NA-60108: Framework document on IN/B-ISDN integration, Issue 2, October 1994.

ETSI DTR NA-60110: ETSI wide workplan for IN/B-ISDN integration, 1994.

ETSI DTR NA-60902: Intelligent network CS-2 – targeted telecommunication services, version 3, March 1993.

ETSI DTR NA-61301: IN/UMTS framework document, 1995.

ETSI DTR NA-61302: IN architecture and functionality for the support of CTM, 1995.

ETSI DR NA-60801: IN management requirements for CS-2, 1994.

ETSI DTR NA-60802: IN management requirements and capabilities for interworking between IN structured networks, 1994.

ETSI DTR NA-60107: Report on long-term evolution, 1994.

ETSI DTR SMG-50104: Scenarios and considerations for the introduction of the universal mobile telecommunications system (UMTS), version 0.4.0, December 1995.

ETSI DTR NA-43308: Baseline document on the integration of IN and TMN, September 1992.

ETSI DTR NA-60301: Interworking between IN-structured networks for CS 2, 1994.

ETSI DTR NA-60401: Enhancements of the IN distributed functional architecture for CS-2, 1994.

ETSI ETR NA-61010: IN user's guide for CS-1, version 07.1, May 1994.

ETSI ETR NA-70201: UPT, general service description, 1992.

ETSI ETS 300348: IN CS-1 physical plane, November 1993.

ETSI ETS 300374-1: IN CS-1 core INAP, part 1: protocol specification, 1994.

ETSI ETS MI NA-60003: NA6 baseline document, version 3, May 1995.

ETSI Group NA6: NA6 work programme, Lisbon, July 1993

ETSI SMG3 TDOC 95C101: The CAMEL feature, service description, stage 1, GSM 02.78, version 0.7.1, June 1995.

ETSI DTR NA-60002: IN vocabulary, 1994.

ETSI DTR NA-60109: Service life cycle reference model, 1994.

ETSI TCR-TR NA-60502: Distributed functional plane for IN CS-1, November 1993.

ETSI TCR-TR NA-60106: Intelligent network: framework, March 1992.

ETSI TCR-TR NA-60204: Guidelines for standards on IN capability set 1, March 1992.

ETSI TCR-TR NA-60501: Global functional plane for IN CS-1, November 1993.

ETSI DTR NA-60605: Global functional plane for CS-2, 1994.

ETSIGSM-PN GSM recommendations, Geneva, September 1989.

ISO/IEC/IS 10165-1/ITU Recommendation X.720: Information processing systems–open system interconnection – structure of management information – parts 1–5, 1992.

ISO/IEC/IS 9595 / ITU Recommendation X.710: Information processing systems–open system interconnection – common management information service definition (CMIS), 1991.

ISO/IEC/IS 9596-1/ITU Recommendation X.711: Information processing systems–open system interconnection – common management information protocol definition (CMIP), 1991.

ITU Draft Recommendation F.851: Universal personal telecommunications – service principles and operational provision, November 1991.

ITU Draft Recommendation Q.1221: Introduction to intelligent network capability set 2, Berlin, Nocember 1995.

ITU Draft Recommendation Q.1222: Intelligent network CS-2 service plane, Berlin, November 1995.

ITU Draft Recommendation Q.1223: Global functional plane for intelligent network CS-2, Berlin, November 1995.

ITU Draft Recommendation Q.1224: Distributed functional plane for intelligent network CS-2, Berlin, November 1995.

ITU Draft Recommendation Q.1225: Physical Plane for Intelligent Networks CS-2, Berlin, November 1995.

ITU Draft Recommendation Q.1228: Intelligent interface tradition for CS-2, Berlin, November 1995.

ITU Draft Recommendation Q.1229: IN Users Giide fr CS-2, Berlin, November 1995.

ITU Draft Recommendation Q.1290: Glossary of terms used in the definition of intelligent networks, Geneva, May 1995.

ITU Recommendation I.130: Method for the characterization of telecommunication services supported by an ISDN and network capabilities of an ISDN, Melbourne, 1988.

ITU Recommendation M.3010: Principles for a telecommunications management network.

ITU Recommendation M.3020: TMN interface specification methodology.

ITU Recommendation M.3200: TMN management services: overview.

ITU Recommendation M.3400: TMN management functions.

ITU Draft Recommendation M.32IN: TMN management services for IN, Geneva, 1995.

ITU Recommendation Q.1200: General series intelligent networks recommendations structure, Geneva, March 1992, revised version approved in May 1995.

ITU Recommendation Q.1201/I.312: Principles of intelligent network architecture, Geneva, March 1992, revised version approved in May 1995.

ITU Recommendation Q.1202/I.328: Intelligent network service plane architecture, Geneva, March 1992, revised version approved in May 1995.

ITU Recommendation Q.1203/I.329: intelligent network global functional plane architecture, Geneva, March 1992, revised version approved in May 1995.

ITU Recommendation Q.1204: Intelligent network distributed functional plane architecture, Geneva, March 1992, revised version approved in May 1995.

ITU Recommendation Q.1205: Intelligent network physical plane architecture, Geneva, March 1992, revised version approved in May 1995.

ITU Recommendation Q.1208: IN interface recommendations – general, Geneva, March 1992, revised version approved in May 1995.

ITU Recommendation Q.1211: Introduction to intelligent network capability set 1, Geneva, March 1992, revised version approved in May 1995.

ITU Recommendation Q.1213: Global functional plane for intelligent network CS-1, Geneva, March 1992, revised version approved in May 1995.

ITU Recommendation Q.1214: Distributed functional plane for intelligent network CS-1, Geneva, March 1992, revised version approved in May 1995.

ITU Recommendation Q.1215: Physical plane for intelligent network CS-1, Geneva, March 1992, revised version approved in May 1995.

ITU Recommendation Q.1218: Intelligent network interface recommendations for CS-1, Geneva, March 1992, revised version approved in May 1995.

ITU Recommendation Q.1219: IN users' guide for CS-1, Geneva, March 1992, revised version approved in May 1995.

ITU Recommendation X.500 / ISO/IEC/IS 9594: Information processing–open systems interconnection – the directory, Geneva, 1988.

ITU Recommendation X.700/ISO/IEC/IS 7498-4: Information processing–open systems interconnection – basic reference model – part 4: management framework, 1989.

ITU Recommendation X.900/ISO/IEC 10746-2.2: Basic reference model for open distributed processing, parts 1–5, June 1993.

ITU Study Group XI: Baseline document for capability sets for the 1993–1996 study period, September 1992.

TINA-C Document. No. TB_A2.HC.012_1.2_94: Computational modeling concepts, December 1994.

TINA-C Document. No. TB_EAC.001_1.2_94: Information modeling concepts, December 1994.

TINA-C Document. No. TB_GN.010_2.0_94: Management architecture, December 1994.

TINA-C Document. No. TB_JJB.005_1.5_94: Connection management architecture, December 1994.

TINA-C Document. No. TB_MDC.018_1.0_94: Overall concepts and principles of TINA, December 1994.

TINA-C Document. No. TB_MH.002_2.0_94: Requirements upon TINA-C architecture, December 1994.

TINA-C Document. No. TB_NS.005_2.0_94: Engineering modeling concepts (DPE architecture), December 199.4

TINA-C Document. No. TP_AJH.001_0.10_94: Deployment scenarios for interworking, December 1994.

TINA-C Document. No. TP_MDC.012_2.0_94: Definition of service architecture, February 1995.

TINA-C Document. No. TR_KMK.001_1.1_94: Engineering modeling concepts (DPE kernel specification), November 1994.

# Index

*Comment:* Note that terms appearing in the text and within a figure on the same page are referenced twice (e.g. see access features).

Page numbers appearing in **bold** refer to figures and page numbers in *italic* refer to tables.